Quintic Equations and How to Solve Them

C. M. Linton

Quintic Equations and How to Solve Them

Birkhäuser

C. M. Linton
Rothley, Leicestershire, UK

ISBN 978-3-032-01657-7 ISBN 978-3-032-01658-4 (eBook)
https://doi.org/10.1007/978-3-032-01658-4

Mathematics Subject Classification: 12-01, 12-03, 12D10, 12F10, 12E05

This book is published under the imprint Birkhäuser, www.birkhauser-science.com by the registered company Springer Nature Switzerland AG
The registered company address is: Gewerbestrasse 11, 6330 Cham, Switzerland

Preface

This book is about the algebraic solution of quintic equations. Such equations hold a special place in the mathematical world because, unlike quadratics, cubics and quartics, there is no formula which allows us to solve the general quintic equation. Equations of degree 2, 3 and 4 are solvable in a way which is not the case for equations of degree 5 or higher.

The story of the quintic and its unsolvability is a well-told tale; it contains fascinating historical characters and beautiful mathematics. But there is one part of this story that is not usually mentioned, and my main reason for writing this book is to rectify this omission. Although there is no general formula for the solution to a quintic equation that uses only a finite number of algebraic operations, some quintic equations *can* be solved algebraically. A lot of mathematical machinery is required in order to find out if a given quintic falls into this category and I have described first how to determine whether a particular quintic can be solved in this way. It is Galois theory that provides the key. This is by no means a trivial process, but it is well known. It is also indirect in that it tells us whether or not an equation can be solved, but not how to solve it.

So, if a given quintic is amenable to an algebraic solution, how do we actually determine the solutions? This problem was unlocked at the end of the nineteenth century, but the solution method has languished in obscurity since then. To my knowledge, there is no exposition of the quintic, designed for a general mathematical audience, that covers this material. Hence this book.

The book is suitable for anyone with an undergraduate mathematics education and, with the exception of the final chapter, the only prerequisite from a technical perspective is a familiarity with high-school algebra. Everything else that is needed is built up from this base. This does not mean, of course, that the material is easy to absorb if it is all new to you. Many of the concepts that are introduced are highly abstract and ones that would not normally be encountered until undergraduate level. For the mathematically curious reader who has not previously encountered Galois theory and/or group theory, there is a lot to assimilate. However, I have only treated those topics that are necessary to be able to identify which quintic equations can be solved algebraically and then how to solve them. There is much, much more that could be said about Galois theory and group theory, and while the fundamentals of these branches of mathematics are explained in this book, there is a richness and depth to these topics that goes way beyond that which is presented here.

Every effort has been made to make the journey navigable for those who have not had the benefit of an undergraduate mathematics education. Lots of historical information is included to provide colour and context, though the book is not primarily a book about the history of mathematics. It is written to try and make the material as accessible as possible, so some things which will be well known to many mathematically knowledgeable readers are explained rather than assumed. Most of the results that are needed are proved in the text, particularly where the proof is either simple or illuminating. On the other hand, some of the longer and more technical proofs are omitted though in that case references are provided for those who want to follow them up. I have chosen [13] as the primary source of many of these proofs, largely because of the book's comprehensive nature. There are of course many other textbooks that cover this material. Numerous examples, mostly explicit calculations for particular polynomials, are included to demonstrate how the theory applies in practice.

The story of the quintic and its unsolvability fills Chaps. 3–7, roughly the first half of the book. Following an Introduction, Chap. 2 describes how to solve quadratic, cubic and quartic equations. This requires working with complex numbers, objects that are necessary to make sense of the algebraic solution to the cubic and higher-order equations. The formal groundwork for a theory of polynomial equations, with coefficients in a given field F, is laid in Chap. 3, and the cube roots and fifth roots of unity are calculated explicitly. The symmetric polynomials in the roots of a polynomial are discussed in Chap. 4, and this leads on to a description of permutation groups. The symmetric groups on 2, 3 and 4 letters are described in some detail, and many of the basic concepts of group theory, including in particular that of a normal subgroup, are introduced. Chapter 5 addresses the question 'what is the smallest field that contains all the roots of a given polynomial?' and leads to the idea of a splitting field. Then in Chap. 6 we develop the subject of Galois theory, at least as it applies to the solutions of polynomial equations, establishing the correspondence between subfields of the splitting field and subgroups of the symmetric group. We show explicitly, how the solutions to quadratic, cubic and quartic equations fit into this formalism.

The remainder of the book is about quintic equations. Chapter 7 establishes the fact that the general quintic cannot be solved using a finite number of algebraic operations, a so-called solution in radicals. Historically, this was established before Galois' revolutionary ideas had been codified into a coherent theory, but it is Galois theory that allows us to classify quintics into those that can be solved by radicals and those that cannot. A general quintic has five independent parameters, and in Chap. 8, we show how transformations can be used to reduce that number down to 3, or 2, or 1. These transformations are by no means straightforward but they involve nothing more elaborate than square and cube roots to accomplish. The most significant material is that which describes how to solve a solvable quintic and that is the subject of Chap. 9. We will describe the algorithm in detail and give lots of examples of its use. Finally, in Chap. 10 we discuss, somewhat superficially and without explaining all the necessary machinery, some other methods for determining the solutions to quintic equations.

Many of the concepts that thread through this book are rather abstract in nature, but the end goal is not. Given a solvable quintic, the algorithm described in Chap. 9 allows us to determine its solutions in the form of finite, algebraic expressions. This juxtaposition of abstract theory with direct calculation is a recurring theme, and I have included numerous examples to illustrate how the underlying ideas manifest in practice. From Chap. 3 onwards, these examples appear in shaded boxes and are numbered to enable cross-referencing. Key results are presented as numbered propositions in boxes so that they are easy to pick out and refer back to. A few important, named theorems get the same treatment. When an important quantity is defined in the text, I have highlighted this in bold font. I have not included exercises for the reader but if, like me, you enhance your understanding through doing, then I would encourage you to work through the examples that are given and fill in the details. Another activity that I would highly recommend is to turn the method for solving a solvable quintic described in Chap. 9 into an automated process using the computer algebra package of your choice.

I have endeavoured to check and double-check the many algebraic expressions that appear in the book. Some errors will no doubt remain and for these I apologise in advance.

Rothley, UK C. M. Linton

Declarations

Competing Interests The authors have no competing interests to declare that are relevant to the content of this manuscript.

Contents

1 Introduction

The quadratic formula is a wonderful thing. When learning algebra for the first time, after observing some neat ways to solve specific quadratic equations (using factorisation or completing the square, for example) along comes an expression that allows you to solve any quadratic equation simply by plugging the coefficients in the quadratic into a formula. For the equation

$$ax^2 + bx + c = 0,$$

we learn that the solutions are given by

$$x = \frac{-b \pm \sqrt{b^2 - 4ac}}{2a}.$$

First and foremost, this formula is extremely useful. If you want to know the solutions to a quadratic equation quickly, whether as an exact expression or numerically, using this formula is as good a method as any. It really isn't worth trying to spot if there is a neater way.

Secondly, it is completely general. The solutions to *any* quadratic are given by this compact expression. Of course, we can't have $a = 0$ since division by zero is not admissable, but then if $a = 0$, we don't actually have a quadratic equation in the first place and the solution $x = -b/c$ is immediate. There is the tricky point that $b^2 - 4ac$ might be negative, in which case our formula asks us to take the square root of a negative quantity. Depending on whether complex numbers have crossed our path or not, we will either take this in our stride or conclude that there are no solutions. But there is nothing lacking in the formula; values of x given by the above expression do satisfy the original equation and no other numbers do.

Thirdly, the formula only involves the basic arithmetic operations of addition, subtraction, multiplication and division, together with a square root. Taking a square root is not a particularly straightforward operation, except in special cases, and

C. M. Linton, *Quintic Equations and How to Solve Them*,
https://doi.org/10.1007/978-3-032-01658-4_1

again, depending on one's outlook one might leave a solution in terms of a number like $\sqrt{2}$, or one might convert this to a numerical approximation 1.414 . . .. Due to its utility, generality and simplicity, the quadratic formula has an air of magic about it. For many who use it, it is easy enough to remember and the number of people who can use it effectively is no doubt rather larger than the number of people who can remember how to derive it.

For those who are mathematically curious, an obvious question to ask is: does a similar formula exist for cubic equations? If you ask this question at school you might be told that there is a formula but it is too complicated to be particularly useful. Indeed, if you are interested in the numerical value of the solutions to a cubic equation, this is a very reasonable response. In that case it is most likely that some form of numerical approximation technique is what you need, not a complicated formula that involves (as the cubic formula does) cube roots of expressions that themselves contain square roots. Thus, as we shall see, the cubic formula does not possess the simplicity of the quadratic formula and, unless one specifically requires an exact expression for the solutions, is unlikely to be useful.

However, the question of its generality forced mathematicians in the sixteenth century to confront complex numbers. In contrast to the quadratic formula, the cubic formula can involve complex numbers even when the solutions to the equation are all real, a sign that there are subtle nuances at play. The inclusion of complex numbers in our mathematical vocabulary allows us to define square roots, cube roots, fourth roots and so on with a degree of consistency that is not otherwise possible and this is a key step in the creation of a general theory of polynomial equations. The derivation of a formula which expresses the solution to cubic equations using the four basic operations of arithmetic and the extraction of roots (in this case a mixture of cube and square roots) has both a fascinating history, which we will briefly describe in Chap. 2, but more importantly was a catalyst for mathematical developments that had profound and far reaching consequences.

Roughly 2000 years passed between the first successful attempts to solve quadratic equations and the publication of a formula for the solution to cubic equations. Quadratic equations can readily be formulated as problems in geometry and so they were easily accommodated within the prevailing paradigm of ancient Greek mathematics. However, it was the development of algebra in the Islamic world that enabled the cubic to be tamed. The nature of the cubic formula was troublesome, however. It forced mathematicians to broaden their horizons as to the forms that algebraic quantities could take. Indeed, the discovery of the cubic formula was arguably more important for the impetus that it gave to new lines of mathematical enquiry than it was as a method for solving cubic equations.

Quartic equations succumbed almost immediately after the secrets of cubic equations were revealed. A fairly simple process can be used to transform a quartic into a cubic (it involves taking square roots) and we can solve the quartic by first solving the cubic and then inverting the transformation. Simplicity is not a word that should be associated with the resulting process although it is still completely general. By the end of the sixteenth century, it was thus well understood that any polynomial equation of degree 2, 3 or 4 could be solved using a finite number of the operations $+, -, \times, \div$, together with root extractions, which collectively we refer to as algebraic operations. Whether you would call the method for determining the solution to a quartic equation a formula or not is debatable. It might be more appropriately described as a sequence of formulas, but that is more a statement about the limitations of our mathematical notation. If we interpret formula to mean a finite sequence of specified algebraic operations, then it is reasonable to describe the cubic and quartic solutions in this way.

But there the sequence ends. There is no algebraic formula which expresses the solutions to the general quintic equation

$$ax^5 + bx^4 + cx^3 + dx^2 + ex + f = 0, \quad a \neq 0,$$

in terms of the coefficients a, b, c, d, e, f. This is a remarkable statement for a whole host of reasons. Perhaps the first question that comes to mind is how can we possibly know? It seems perfectly reasonable to say that no-one has ever found such a formula, but how can we be certain that no such formula is out there waiting to be discovered. What is special about the number 5? What is it that we can do when the degree is 4 or less that becomes impossible when it is increased to 5? It turns out that the answer concerns the interrelations that exist within collections of five objects, rather than the number 5 itself, and this is where group theory plays a leading role. The theory of groups is now a major branch of mathematics and finds applications in many branches of science, notably physics and chemistry, but it was originally developed specifically to better understand the roots of equations.

The key to understanding that there is no general formula lies in a completely different way of looking at equations and their solutions; we need to explore the relationships between the solutions themselves. It is through understanding how the different solutions to a quintic equation relate to each other that we will be able to show that no general formula to express them in terms of the coefficients of the equation can possibly exist. Over 250 years elapsed between the discovery of the algebraic solution to a general quartic equation and the realisation that the hunt for a general quintic formula was futile. But the search itself led mathematicians to think about equations in new ways and this ultimately led to the creation of a great deal of new and powerful mathematics, including Galois theory and group theory, the concept of a group being one of the most powerful mathematical ideas ever conceived.

There is a parallel here with Fermat's last theorem, arguably the most famous mathematical statement that postulates the non-existence of something. Fermat claimed that if n is an integer that is greater than 2, then there are no integers x, y, z such that $x^n + y^n = z^n$. It took over 350 years for mathematicians to eventually prove Fermat's conjecture and a great deal of new mathematics was needed to unlock the secrets contained within this simple looking equation. The unsolvability of the quintic is nowhere near as difficult to prove as Fermat's last theorem, but both spawned whole new branches of mathematics.

The absence of a general algebraic formula does not mean that quintic equations cannot be solved. If we insert particular values for the coefficients then we can use a numerical algorithm to find approximations to the solutions to any desired level of accuracy. Moreover, in some cases exact answers will be available: the equation $x^5 - 1 = 0$ clearly has $x = 1$ as a solution, for example. A less trivial example is that $\sqrt[5]{4} - \sqrt[5]{2}$ is a solution to the equation $x^5 + 10x^2 + 10x - 2 = 0$. The following question arises, therefore: if we ascertain that a particular quintic equation can be solved algebraically, how do we actually determine the solutions? Finding the answer to this question is our goal.

2 Quadratics, Cubics, Quartics and Complex Numbers

How might we find two numbers which when added together give 10 and when multiplied give 21? Apart from simply spotting the answer, a familiarity with basic algebra might suggest that we denote the two numbers by x and y and then the information given in the question is

$$x + y = 10, \quad xy = 21.$$

We won't be overly concerned in this book with describing practical reasons as to why one might want to do this but, as a simple example, if we wanted to cordon off a rectangular area of 21 m^2 with a tape of length 20 m, then these are the equations one would have to solve for the lengths of the sides. To make progress, we can eliminate y. From the first equation, $y = 10 - x$ and substitution of this expression for y into the second equation yields a quadratic equation for x:

$$x^2 - 10x + 21 = 0. \tag{2.1}$$

If we posess a general method for solving quadratic equations then it can be used to show that x can be either 7 or 3, with y then the other of these two numbers. The power in this algebraic method lies not in enabling one to solve this particular problem, but in allowing any problem of this type to be solved. By repeating the above argument with 10 and 21 replaced by $-b$ and c, respectively, we can see that asking for two numbers which add together to give $-b$ and when multiplied give c, with b and c being arbitrary numbers, is equivalent to asking for a solution to the quadratic equation

$$x^2 + bx + c = 0. \tag{2.2}$$

Quadratic equations have been studied for thousands of years, with examples of solutions to specific equations documented on Babylonian cuneiform tablets dated

C. M. Linton, *Quintic Equations and How to Solve Them*,
https://doi.org/10.1007/978-3-032-01658-4_2

around 2000–1500BC and similar problems appearing in Chinese texts from about 1000 years later. Euclid's *Elements*, written in about 300BC, contains what are essentially geometric solutions to quadratic equations and algebraic approaches to the solution of quadratics are evident in the work of Diophantus in the third century AD. However, it wasn't until the work of the Islamic scholars of the ninth century that general methods were clearly articulated.

Islamic mathematicians built on the body of work left by their predecessors in Babylonia, Greece, Persia, India and, to some extent, China. Works were translated into Arabic, including the creation of new technical vocabulary, which greatly facilitated the assimilation of all this knowledge and allowed others to forge ahead with new ideas. During the course of the eighth and ninth centuries, scholars created a centre of learning in Baghdad, known as the *Bayt al-Ḥikma* (House of Wisdom), and it was through the work of one of those scholars, Muḥammad ibn Mūsā al-Khwārizmī, that Arabic mathematics really came of age.

His *Ḥisāb al-jabr wa-l-muqābala*, written in about 825, contains the beginnings of the subject now known as algebra, *al-jabr* referring to the operation of taking a subtracted quantity from one side of an equation to become an added quantity on the other, for example, in modern algebraic language,

$$x - a = y \quad \text{implies} \quad x = y + a,$$

and *al-muqābala* referring to the operation of subtracting the same quantity from both sides of an equation, as in

$$x + a = y + a \quad \text{implies} \quad x = y.$$

Al-Khwārizmī is also responsible for introducing the Hindu decimal numeration system, including a symbol for zero, to the Arabic speaking world. He described algorithms (the word being derived from a Latinized version of his name) for using these numbers to perform the basic operations of arithmetic: addition, subtraction, multiplication and division.

What al-Khwārizmī did not have access to was the algebraic symbolism that we use today. All his methods were described using specific problems, but they were written very clearly as general methods which could easily be adapted to problems of a similar type. Prior to the sixteenth century only the Greek mathematician Diophantus had made a conscious effort to introduce symbolism to facilitate algebraic thinking. Symbols like $+$ and $-$ started to be introduced in the fifteenth century and the equals sign was introduced by Robert Recorde in 1557. The introduction of letters to represent unknown or arbitrary quantities was developed in the sixteenth century, pioneered by people such as François Viète, and the modern practice of using letters toward the end of the roman alphabet x, y, z to represent unknowns and those at the beginning a, b, c to represent fixed but arbitrary parameters is usually credited to René Descartes, who also introduced the $\sqrt{\ }$ symbol for extracting roots.

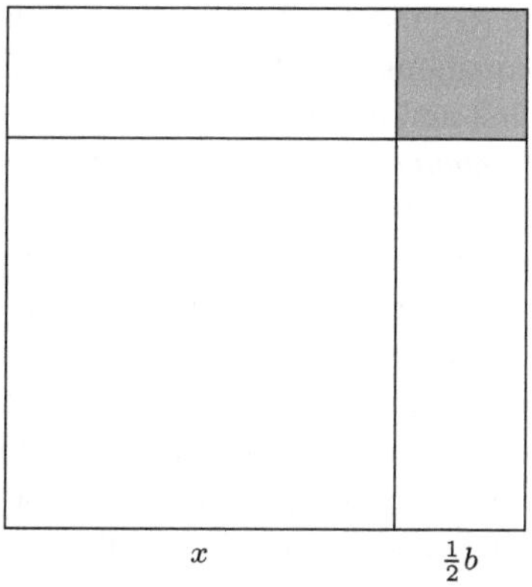

Fig. 2.1 Completing the square. The complete square is made up of a large square with area x^2, two rectangles which are $b/2 \times x$ in size (and thus bx in total) and a small square (shaded) of size $(b/2)^2$. Thus $x^2 + bx = (x + b/2)^2 - b^2/4$

Although he did not have the symbolism to enable him to express it, al-Khwārizmī had an algebraic method that was equivalent to the technique now referred to as 'completing the square' and he illustrated his methods with diagrams which show clearly that he was using a geometrical analogy; see Fig. 2.1. Since, from the diagram,

$$x^2 + bx = \left(x + \tfrac{1}{2}b\right)^2 - \tfrac{1}{4}b^2,$$

we can rewrite the quadratic equation (2.2) as

$$\left(x + \tfrac{1}{2}b\right)^2 = \tfrac{1}{4}b^2 - c.$$

This enables us to take the square root (noting that there are potentially two) and then rearrange the resulting expression to give the quadratic formula

$$x = \tfrac{1}{2}(-b \pm \sqrt{b^2 - 4c}). \tag{2.3}$$

One of the curiosities of quadratic equations, completely understood by al-Khwārizmī, was that sometimes they would have two solutions as in the case of eq. (2.1), sometimes one, and sometimes none at all. He also knew that the sign of the quantity $b^2 - 4c$ determined the number of roots, the number of roots being 2, 1, or 0, depending on whether $b^2 - 4c$ is greater than, equal to, or less than, zero. The formula (2.3) solves the quadratic equation (2.2) in which the coefficient of x^2 is one. If we are faced with the equation $ax^2 + bx + c = 0$, with non-zero a, then we can simply divide through by a first.

Once the idea of a general solution to quadratic equations is understood, it is natural (for mathematicians!) to try and generalise this idea and see if a similar result can be obtained for equations containing higher powers of x. In the fifteenth and early sixteenth centuries, as the study of mathematics was beginning to flourish in Renaissance Europe, a number of mathematicians put their minds to solving cubic equations, equations where the highest power of x is 3. Little progress was made at first and not everyone was convinced that this was a fruitful line of enquiry. In his *Summa de arithmetica* (1494), a summary of most of the known mathematics of the time, Luca Pacioli, an Italian mathematician and Franciscan friar, asserted that no general solution to the cubic was possible.

Algebra was still in its infancy and at that time coefficients in an equation were usually taken to be positive, so a cubic equation such as $x^3 + px + q = 0$ was considered different to one of the form $x^3 + px = q$. This leads to lots of different cases, each of which needs to considered independently. A method that works for one type of equation might not work for another. Many attempted to rise to the challenge and in the end the secrets of the cubic were revealed, but not without a great deal of controversy.

The discovery of the algebraic solution of cubic equations is one of the great tales in the history of mathematics, involving a cast of colourful characters and laced with plenty of intrigue, to which we cannot do justice here. For an entertaining and thorough account, see [68]. A brief synopsis goes something like this. The first person to develop an algebraic solution (for cubics of the form $x^3 + px = q$) was a professor at the University of Bologna, Scipione del Ferro, sometime in the early sixteenth century. However, he kept his solution to himself hoping that it would help him in winning a potential mathematical challenge—a common part of academic life at the time. As well as being a blow to one's reputation, losing such a challenge might cost you your job, so having a few aces up your sleeve was a sensible insurance policy. Before he died, in 1526, he passed on his discoveries to one of his pupils, Antonio Maria Fior.

A few years' later, and completely independently, Niccolò Tartaglia discovered a general method for solving $x^3 + px = q$. Tartaglia was a respected mathematician and engineer who taught practical mathematics and bookkeeping, from 1534 until his death in 1557 in the Republic of Venice. Tartaglia did let it be known that he was in possession of a general method for these types of cubic equation but, just like del Ferro, he kept his solution to himself. In 1535 Fior, perhaps thinking that Tartaglia was bluffing, decided to challenge him to a competition. Fior's confidence was misplaced and Tartaglia prevailed easily.

News of Tartaglia's success spread rapidly within scientific circles in northern Italy and soon reached Gerolamo Cardano in Milan. Cardano was an esteemed doctor and talented mathematician, with friends in high places, and he invited Tartaglia to visit him. Tartaglia made the journey to Milan in 1539 and was persuaded by Cardano's flattery to divulge his method. In return, Cardano took an oath not to publish the solution so as to give Tartaglia the time to make it public himself. Tartaglia was busy translating Euclid's *Elements* into Italian at the time, but does not seem to have been in any hurry to get his algebraic discovery into print.

Once he had understood the principles at the heart of Tartaglia's approach, Cardano and his pupil Lodovico Ferrari then set to work trying to understand the algebra of cubic equations and managed to derive a complete solution. Indeed, the brilliant Ferrari solved quartic equations at the same time by reducing them to cubic equations. When Cardano subsequently found out that the original discovery was made by del Ferro and not by Tartaglia he decided that he had allowed plenty of time for Tartaglia to publish his work and that his obligation to him had ended. He thus presented the complete solution to cubic and quartic equations in his *Ars Magna* of 1545, heaping praise on Ferrari. Even though Cardano did give Tartaglia credit for his independent discovery, Tartaglia was furious and he was very public in his scathing criticism of Cardano. Ferrari was keen to defend his teacher's honour and this ultimately led to another public contest, this time between Tartaglia and Ferrari, and Ferrari's much better grasp of the subject meant he came out on top. To the victor, the spoils: the algebraic solution for the cubic is known as Cardano's formula, with Ferrari credited with the solution to the quartic. Del Ferro's and Tartaglia's contributions are left to the history books.

Before describing Cardano's solution to the cubic, we note that it is straightforward to remove the x^2 term, if it is present, from a general cubic. Given the equation

$$x^3 + bx^2 + cx + d = 0,$$

then we can apply the transformation $x = y - b/3$. Substitution of this into the general cubic then shows, with the aid of the binomial expansion

$$(a + b)^3 = a^3 + 3a^2b + 3ab^2 + b^3, \tag{2.4}$$

that y satisfies the cubic

$$y^3 + py + q = 0, \tag{2.5}$$

with

$$p = c - \tfrac{1}{3}b^2, \quad q = d + \tfrac{2}{27}b^3 - \tfrac{1}{3}bc.$$

An equation of the form (2.5), in which the term in the square of the unknown is absent, is called a **depressed cubic**. If we can solve this for y, then the roots of the original cubic are found simply by subtracting $b/3$.

To derive Cardano's formula, let $y = u + v$ and substitute this into (2.5) to obtain

$$u^3 + v^3 + (3uv + p)(u + v) + q = 0.$$

This equation will be satisfied if we have both

$$u^3 + v^3 = -q \quad \text{and} \quad 3uv = -p.$$

Thus u^3 and v^3 are such that their sum is $-q$ and their product is $-p^3/27$, which is exactly the type of problem we posed right at the beginning of the chapter and leads to the solution of a quadratic equation. In this case u^3 and v^3 are solutions to $x^2 + qx - p^3/27 = 0$ and hence, using the quadratic formula, we find that

$$u^3 = -\tfrac{1}{2}q + \sqrt{D}, \quad v^3 = -\tfrac{1}{2}q - \sqrt{D}, \quad \text{where} \quad D = \tfrac{1}{27}p^3 + \tfrac{1}{4}q^2.$$

We can then write (Cardano's formula)

$$y = \sqrt[3]{-\tfrac{1}{2}q + \sqrt{D}} + \sqrt[3]{-\tfrac{1}{2}q - \sqrt{D}}, \tag{2.6}$$

a neat, algebraic expression for the solution to a cubic equation.

Here we should take stock and reflect on what the symbols $\sqrt{\ }$ and $\sqrt[3]{\ }$ actually mean, at least in the context of positive and negative numbers. For square roots, if x is a positive quantity, $\sqrt{x}$ always means the positive square root; thus $\sqrt{4} = 2$. The two solutions to the equation $x^2 = 4$ are then represented as $\pm\sqrt{4} = \pm 2$. For cube roots, we again have that $\sqrt[3]{x}$ is positive if x is; $\sqrt[3]{8} = 2$, for example. But we can also define $\sqrt[3]{x}$ for negative x and the answer will be negative: $\sqrt[3]{-8} = -2$, since $(-2)^3 = -8$. This way of looking at cube roots is not sufficiently general for our needs but for the time being, it is all we have.

So what does Cardano's formula appear to imply? If $D < 0$ (which can happen if p is negative) we immediately have a problem, since the expressions for u^3 and v^3 involve the square root of a negative quantity. As in the case of quadratic equations we might therefore conclude that there simply isn't a solution in this case. On the other hand, if $D \geq 0$, then (2.6) results in a unique solution.

Let's look at an example where everything works well. Consider

$$y^3 + 3y + 2 = 0,$$

for which Cardano's formula applies with $q = D = 2$. Hence

$$y = \sqrt[3]{-1 + \sqrt{2}} + \sqrt[3]{-1 - \sqrt{2}}. \tag{2.7}$$

This is a unique, well-defined number (according to the convention on the use of the symbol $\sqrt[3]{\ }$ described above) and is indeed a solution to the given cubic equation, as can easily be checked.

An example where things work, but somewhat less well, is

$$y^3 + 6y - 20 = 0.$$

Since $y^3 + 6y - 20 = (y - 2)(y^2 + 2y + 10)$, we deduce that 2 is the only real root of the cubic but an application of Cardano's formula leads to the result

$$y = \sqrt[3]{10 + 6\sqrt{3}} + \sqrt[3]{10 - 6\sqrt{3}}.$$

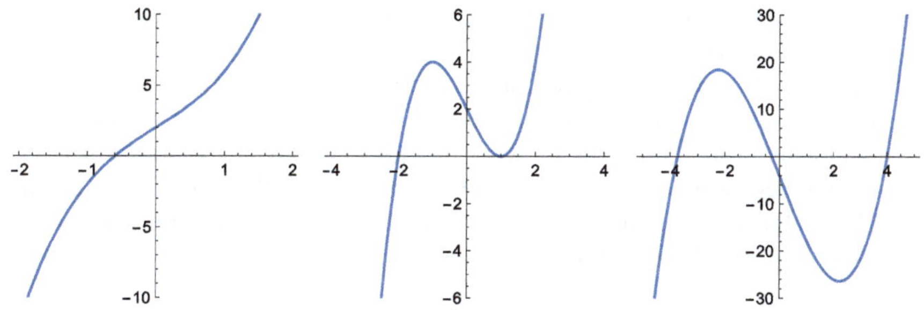

Fig. 2.2 Three cubic curves. From left to right, $y^3 + 3y + 2$, $y^3 - 3y + 2$, $y^3 - 15y - 4$

Is this equal to 2? The answer is yes, but it is far from obvious. Using the binomial theorem, we can show that $(1 \pm \sqrt{3})^3 = 10 \pm 6\sqrt{3}$. Hence the cube roots can be extracted and $y = 1 + \sqrt{3} + 1 - \sqrt{3} = 2$.

If we plot the curve $y^3 + 3y + 2$, as is shown in Fig. 2.2, we get a curve which has a single zero, which is given by Cardano's formula in (2.7). But cubics can have one, two or three roots, as the other examples in Fig. 2.2 illustrate. The second example is $y^3 - 3y + 2 = 0$, for which $D = 0$. Here Cardano's formula leads to the solution -2, but the solution 1 is missed. And the third example is even more disturbing since $y^3 - 15y - 4 = 0$ has three solutions but since $D < 0$ in this case, Cardano's formula does not appear to give any of them. So what is going on?

Let us look at the equation

$$y^3 - 15y - 4 = 0$$

in more detail. This is one of the equations that was discussed by the engineer Rafael Bombelli in his *Algebra* of 1572, a work written in Italian rather than Latin and designed to make the recent developments in algebra more accessible. For this equation, $p = -15$, $q = -4$ and hence $D = -121$, leading to the formula

$$y = \sqrt[3]{2 + \sqrt{-121}} + \sqrt[3]{2 - \sqrt{-121}}. \tag{2.8}$$

However, as Bombelli noted, a brief inspection of the original equation shows that $y = 4$ is a solution. Rather than discarding Cardano's formula, Bombelli took the radical step of treating the square roots of negative numbers as algebraic quantities and manipulating them using the standard operations of arithmetic. Thus, using (2.4), he computed

$$(2 \pm \sqrt{-1})^3 = 2 \pm 11\sqrt{-1} = 2 \pm \sqrt{-121}.$$

This allows us to extract the cube roots in (2.8) and we obatin $y = 2 + \sqrt{-1} + 2 - \sqrt{-1} = 4$. This is clearly not the whole story as $y^3 - 15y - 4 = 0$ has three solutions and we have only found one using Cardano's formula, but it does show that there is information locked in that formula that requires a new way of thinking about square roots and cube roots.

What Bombelli, and to a degree Cardano before him, was doing was introducing into mathematics the concept of complex numbers. It took quite a while before mathematicians were comfortable with the idea, hardly surprising since even negative numbers were still something of a mystery, but they are fundamental to the study of algebraic equations. Putting the historical development to one side, we take a bold step and simply define the quantity

$$\mathrm{i} = \sqrt{-1}.$$

In other words i has the property that if we square it we get -1. This is not a number in the usual sense of the word (in other words, not a real number), since it has a negative square, but we choose to create such a thing and see where it takes us. Apart from the defining property $\mathrm{i} = \sqrt{-1}$, we are going to insist that i satisfies all the normal rules of arithmetic and so, in particular, $(-\mathrm{i})^2 = (-1)^2\mathrm{i}^2 = \mathrm{i}^2$ and so $-\mathrm{i}$ is also a square root of -1.

In general, we can consider **complex numbers** of the form $x + \mathrm{i}y$, where x and y are real numbers, and then calculate

$$(a + \mathrm{i}b) + (c + \mathrm{i}d) = a + c + \mathrm{i}(b + d),$$
$$(a + \mathrm{i}b)(c + \mathrm{i}d) = ac - bd + \mathrm{i}(ad + bc),$$
$$\frac{a + \mathrm{i}b}{c + \mathrm{i}d} = \frac{(a + \mathrm{i}b)(c - \mathrm{i}d)}{c^2 + d^2} = \frac{ac + bd}{c^2 + d^2} + \mathrm{i}\frac{bc - ad}{c^2 + d^2},$$

which shows that if we add, subtract, multiply or divide complex numbers, excepting division by zero, the result is always a complex number. If $z = x + \mathrm{i}y$, then we call x the **real part** of z (written $x = \mathrm{Re}\, z$) and y the **imaginary part** (written $y = \mathrm{Im}\, z$). Also, the number $x - \mathrm{i}y$ is the **complex conjugate** of z and we denote this by $\overline{z}$. The **modulus** of a complex number, $|z|$, is defined via

$$|z|^2 = z\overline{z} = x^2 + y^2.$$

Armed with these new algebraic quantities, taking square roots of negative quantities is now straightforward; we simply have $\sqrt{-a} = a\mathrm{i}$. If we look back at the quadratic formula, we can express the solutions of $x^2 + bx + c = 0$, in the case when $b^2 - 4c < 0$, as

$$x = \tfrac{1}{2}(-b \pm \mathrm{i}\sqrt{4c - b^2}).$$

These quantities satisfy the equation, in that if you evaluate $x^2 + bx + c$ using the rules of algebra togther with $\mathrm{i}^2 = -1$ the result is zero, but it is not immediately apparent what is to be gained from this insight. As Cardano wrote: "So progresses arithmetic subtlety the end of which ... is as refined as it is useless"! The sense of mystery surrounding complex numbers, even to those who were comfortable manipulating them formally, is illustrated through the words of the great polymath Gottfried Leibniz, written about 100 years after Bombelli's pioneering work [38, Chapter 13]:

> The Divine Spirit found a sublime outlet in that wonder of analysis, that portent of the ideal world, that amphibian between being and non-being, which we call the imaginary root of negative unity.

Leibniz's early musings on complex numbers are described in [45].

Taking cube roots is not so straightforward. We begin by trying to take the cube root of 1. This is equivalent to solving the cubic equation $x^3 - 1 = 0$. Since we know that 1 is a solution, we can factorise this to yield

$$x^3 - 1 = (x-1)(x^2 + x + 1) = 0$$

and the solutions to $x^2 + x + 1 = 0$ are therefore also cube roots of 1. The quadratic formula shows that these two additional solutions (if we admit complex numbers) are given by $(-1 + \mathrm{i}\sqrt{3})/2$ and $(-1 - \mathrm{i}\sqrt{3})/2$. If we label the first of these two complex roots as ω, then either by direct calculation or by looking at the equation that they both satisfy, we find that the other solution is ω^2. Hence the three roots of $x^3 - 1 = 0$ are $1, \omega, \omega^2$, where

$$\omega = \tfrac{1}{2}(-1 + \mathrm{i}\sqrt{3}), \quad \omega^2 = \tfrac{1}{2}(-1 - \mathrm{i}\sqrt{3}) = \overline{\omega}. \tag{2.9}$$

Now consider the equation $x^3 = a$, where a is any real or complex number and let's assume that u is a (real or complex) number such that $u^3 = a$. Then $x^3 = a$ becomes $x^3 = u^3$ and we can take the cube root to give $x = u\sqrt[3]{1}$. Thus there are three cube roots of a, given by u, ωu and $\omega^2 u$. We are now in a position to provide a better definition of cube root. For any a, $u = \sqrt[3]{a}$ signifies a number for which $u^3 = a$. When a is real and positive, $\sqrt[3]{a}$ is taken to be real and positive, but apart from in this special case, no particular root is indicated by $\sqrt[3]{a}$. In particular, if a is real and negative we do not assume that $\sqrt[3]{a}$ is real and negative. For any a, there are then three cube roots: u, ωu and $\omega^2 u$.

Let us now return to Cardano's formula for the solution to $y^3 + px + q = 0$:

$$y = \sqrt[3]{-\tfrac{1}{2}q + \sqrt{D}} + \sqrt[3]{-\tfrac{1}{2}q - \sqrt{D}}, \quad D = \tfrac{1}{27}p^3 + \tfrac{1}{4}q^2.$$

Each cube root now takes three possible values, which would appear to suggest 9 possible values for y. But if we look back at the derivation we see that the cube roots must be chosen so that their product is $-p/3$. Once we have found a cube root, u, the other cube root is $v = -p/3u$, and then, since $\omega^3 = 1$, the three solutions to $y^3 + px + q = 0$ are given by

$$y_1 = u + v, \quad y_2 = \omega^2 u + \omega v, \quad y_3 = \omega u + \omega^2 v. \tag{2.10}$$

If we go back to the equation $y^3 - 15y - 4 = 0$ considered by Bombelli, with solutions given by (2.8), we need to take cube roots of $2 \pm \sqrt{-121}$ such that their product is 5. We have already extracted the cube roots $2\pm\mathrm{i}$ and $(2+\mathrm{i})(2-\mathrm{i}) = 2^2 - \mathrm{i}^2$ is indeed 5; hence the three solutions can be calculated as

$$\begin{aligned}
y_1 &= 2 + \mathrm{i} + 2 - \mathrm{i} = 4,\\
y_2 &= \omega^2(2 + \mathrm{i}) + \omega(2 - \mathrm{i}) = -2 + \sqrt{3},\\
y_3 &= \omega(2 + \mathrm{i}) + \omega^2(2 - \mathrm{i}) = -2 - \sqrt{3},
\end{aligned}$$

where we need to use the expressions for ω and ω^2 given in (2.9). The solutions y_2 and y_3 could have been obtained much more easily by using the fact that 4 is a solution to factorise the equation into

$$y^3 - 15y - 4 = (y - 4)(y^2 + 4y + 1) = 0.$$

Then solving the quadratic $y^2 + 4y + 1 = 0$ leads to the solutions $-2 \pm \sqrt{3}$. But that is to miss the point. Cardano's formula is an algebraic expression which yields the solution to a general cubic equation. Remarkably, it seems to necessitate the introduction of square roots of negative quantities into the world of algebra, and this perhaps suggests that there are some deep ideas at play that we have only just scratched the surface of. This is indeed the case and these ideas are the subject of the remaining chapters of this book.

What of quartic equations? We will have much more to say about them later, but let us describe briefly Ferrari's approach and then an alternative devised by Descartes. Ferrari's insight was to attempt to write a general quartic

$$f(x) = x^4 + bx^3 + cx^2 + dx + e \tag{2.11}$$

as the difference of two squares:

$$f(x) = (x^2 + Ax + B)^2 - (Cx + D)^2. \tag{2.12}$$

If we can find A, B, C, D in terms of b, c, d, e then we can use the identity $p^2 - q^2 = (p + q)(p - q)$ to write this expression as

$$f(x) = \left(x^2 + (A + C)x + (B + D)\right)\left(x^2 + (A - C)x + (B - D)\right)$$

and the four solutions to $f(x) = 0$ come from solving two quadratics.

Expanding the right-hand side of (2.12) and equating coefficients of the different powers of x with those in (2.11) leads to expressions for A, B, C, D in terms of b, c, d, e. We find that $A = b/2$ and we can eliminate C and D to yield a cubic equation for $y = 2B$:

$$y^3 - cy^2 + (bd - 4e)y + 4ce - b^2e - d^2 = 0. \tag{2.13}$$

If $y = y_1$ is a solution of this cubic, then C and D can be determined from

$$C^2 = \tfrac{1}{4}b^2 - c + y_1, \quad D^2 = \tfrac{1}{4}y_1^2 - e, \quad 2CD = \tfrac{1}{2}by_1 - d. \tag{2.14}$$

The four roots of the original quartic are the two roots of each of the two quadratic equations (with both plus signs or both minus signs)

$$x^2 + \left(\tfrac{1}{2}b \pm C\right)x + \tfrac{1}{2}y_1 \pm D = 0. \tag{2.15}$$

To summarise, to solve a quartic you need first to find a root of a cubic, (2.13), then extract two square roots, for C and D as in (2.14), and finally solve two quadratic equations (which, due to the quadratic formula, is just more square roots).

An alternative approach, devised later by Descartes, runs as follows. We start with a depressed quartic, i.e. one with no cubic term,

$$y^4 + py^2 + qy + r = 0.$$

A general quartic $x^4 + bx^3 + cx^2 + dx + e$ can be reduced to one of this type via the transformation $x = y - b/4$. We can also assume that $q \neq 0$, otherwise we are dealing with a quadratic equation for y^2 (known as a biquadratic) which can be solved using the quadratic formula and then extracting a square root. Descartes approach was to write the depressed quartic as a product of two quadratic factors:

$$y^4 + py^2 + qy + r = (y^2 + sy + \xi)(y^2 - sy + \eta). \tag{2.16}$$

The presence of s in the first factor and $-s$ in the second ensures that there is no cubic term when the factors are multiplied together. Equating coefficients of powers of y shows that we need

$$p = \xi + \eta - s^2, \quad q = s(\eta - \xi), \quad r = \xi\eta.$$

Since we have assumed that $q \neq 0$ we see that s is non-zero. Hence we can write

$$\xi = \tfrac{1}{2}\left(p + s^2 - \frac{q}{s}\right), \quad \eta = \tfrac{1}{2}\left(p + s^2 + \frac{q}{s}\right) \tag{2.17}$$

and s needs to be determined from

$$s^6 + 2ps^4 + (p^2 - 4r)s^2 - q^2 = 0 \tag{2.18}$$

which is a cubic equation for s^2. We can take any of the 6 roots for s, compute ξ and η from (2.17) and then obtain the roots of the depressed quartic by solving the quadratic equations that result when the factors on the right-hand side of (2.16) are set to zero. So although the calculations are different, we still have to solve a cubic equation and then extract square roots.

An example is in order, though we have chosen one where the numbers work out nicely as things can get very complicated otherwise. Consider

$$x^4 + 4x^3 + 2 = 0,$$

for which the cubic that arises in Ferrari's method is

$$y^3 - 8y - 32 = 0.$$

This has $y = 4$ as one solution. We thus need to find C and D such that $C^2 = 8$, $D^2 = 2$ and $CD = 4$, i.e. $C = 2\sqrt{2}$, $D = \sqrt{2}$. This leads to the pair of quadratic equations

$$x^2 + 2(1 - \sqrt{2})x + (2 - \sqrt{2}) = 0,$$
$$x^2 + 2(1 + \sqrt{2})x + (2 + \sqrt{2}) = 0,$$

with solutions

$$\sqrt{2} - 1 \pm \mathrm{i}\sqrt{\sqrt{2} - 1}, \quad -\sqrt{2} - 1 \pm \sqrt{1 + \sqrt{2}}.$$

For Descartes' method we let $y = x + 1$ to yield the depressed quartic

$$y^4 - 6y^2 + 8y - 1 = 0.$$

This leads to the following cubic (for s^2):

$$s^6 - 12s^4 + 40s^2 - 64 = 0.$$

Since $s^2 = 8$ is a solution, we can take $s = 2\sqrt{2}$ and obtain the pair of quadratic equations

$$y^2 + 2\sqrt{2}\,y + (1 - \sqrt{2}) = 0,$$
$$y^2 - 2\sqrt{2}\,y + (1 + \sqrt{2}) = 0.$$

The four solutions of these two equations are

$$-\sqrt{2} \pm \sqrt{1+\sqrt{2}}, \quad \sqrt{2} \pm \mathrm{i}\sqrt{\sqrt{2}-1}$$

and so we obtain the same answers for $x = y - 1$ as before.

So far we have looked at quadratic, cubic and quartic equations. Quadratics are solved by taking a square root. Cubics are solved by taking a square root and two cube roots, though complex numbers are necessary to make the algebraic formulas work. Finally quartics are handled by solving a cubic, with a number of additional square roots needed. The natural hope is that when it comes to quintics we will see something similar, perhaps rather more complicated, perhaps involving fifth roots. This was indeed the assumption of many of the great mathematicians of the seventeenth and eighteenth centuries, though they tried in vain to find the right path through the algebraic labyrinth. It turns out that their goal was illusory, but to understand why we need to step right back to the beginning and develop a whole new way of looking at these equations.

3 Polynomials and Their Roots

In order to cast new light on quadratics, cubics, quartics and then higher-order equations we will need to introduce some nomenclature and definitions. We will work with some standard sets of numbers, getting progressively more general, starting with the set of integers, $\mathbb{Z}$, i.e.

$$\mathbb{Z} = \{\cdots, -2, -1, 0, 1, 2, \cdots\}.$$

Numbers which can be expressed as the ratio of two integers, such as $3/4$ or $-1/2$ are known as rational numbers and the set of all rational numbers is denoted by the symbol $\mathbb{Q}$. Integers are rational numbers because, for example, $3 = 3/1$. There are numbers, such as $\sqrt{2}$ and π, that can be represented via an infinite decimal expansion but which are not elements of the set $\mathbb{Q}$. The fact that $\sqrt{2}$ cannot be written as p/q where p and q are integers was known to the ancient Greeks, whereas it was Johann Lambert in 1761 who proved the same result for π.

Any number which can be represented via a finite or infinite decimal expansion is known as a real number and the set of all real numbers is denoted by the symbol $\mathbb{R}$. Every rational number is a real number with either a finite or a recurring decimal expansion. In the previous chapter we introduced the complex numbers; numbers of the form $a + \mathrm{i}b$, where a and b are real numbers (in other words they are elements of the set $\mathbb{R}$, written $a, b \in \mathbb{R}$). All real numbers are complex numbers (just set $b = 0$ in $a + \mathrm{i}b$) and when we refer to complex numbers in general we mean any complex number, which includes real numbers. The set of all complex numbers is denoted by the symbol $\mathbb{C}$. If all elements of a set A are contained in a set B then we say that A is a **subset** of B, written $A \subset B$. We thus have

$$\mathbb{Z} \subset \mathbb{Q} \subset \mathbb{R} \subset \mathbb{C}.$$

C. M. Linton, *Quintic Equations and How to Solve Them*,
https://doi.org/10.1007/978-3-032-01658-4_3

3.1 Polynomials and Fields

Expressions of the form x^2+bx+c or x^4+px^2+qx+r are examples of polynomials which, in general, have the form

$$a_n x^n + a_{n-1}x^{n-1} + \cdots + a_1 x + a_0, \qquad a_n \neq 0.$$

Here $a_0, \ldots, a_n$ are coefficients and x is an indeterminate variable. If we denote this general polynomial by f, then by definition f has **degree** n, which we write as $\deg f = n$. Thus linear polynomials $(ax+b)$ have degree 1, quadratic polynomials (ax^2+bx+c) are polynomials of degree 2, cubics (ax^3+bx^2+cx+d) have degree 3, and so on. A polynomial of degree 0 is a non-zero constant.[1]

What can be said about the coefficients? In all the examples in Chap. 2 the coefficients were integers, but this is merely for clarity of exposition. Indeed, restricting coefficients to the set of integers $\mathbb{Z}$ is problematic because if we divide two integers we are not guaranteed to get an integer as the answer; this would lead to problems in what follows. We would like to ensure that we can always add, subtract, multiply and divide coefficients with a result that would equally well serve as a coefficient. If we start from $\mathbb{Z}$, then in order to enlarge this set so that division does not force us outside the set we need to consider all the rational numbers $\mathbb{Q}$. A set of numbers with the property that if we apply one of the operations $+, -, \times, \div$ to two elements of the set, division by zero being excluded, we obtain a number which is in the set, is said to be closed under the operations of arithmetic and is called a **field**. Thus $\mathbb{Q}$ is a field, whereas $\mathbb{Z}$ is not.

The set of all real numbers $\mathbb{R}$ is also a field and we showed in the previous chapter that the result of adding, subtracting, multiplying or dividing two complex numbers (excluding division by zero) is always another complex number; hence $\mathbb{C}$ is a field as well. Since $\mathbb{Q}$ is itself a field which is a subset of another field $\mathbb{R}$, we say that $\mathbb{Q}$ is a **subfield** of $\mathbb{R}$. It is also a subfield of $\mathbb{C}$, as is $\mathbb{R}$. In this book, we will only be concerned with fields which contain the rational numbers $\mathbb{Q}$ and that are subsets of $\mathbb{C}$. Thus, when we refer to a field F we always mean a field F with $\mathbb{Q} \subseteq F \subseteq \mathbb{C}$.[2] The notation $A \subseteq B$ signifies that A is a subset of B, including the possibility that $A = B$. If we write $A \subset B$ then we say that A is a **proper subset** of B, meaning that all elements of A are in B but there are some elements of B that are not in A. Similarly, when comparing fields, $F \subset \mathbb{C}$ means that F is a **proper subfield** of $\mathbb{C}$.

For a given field F, we denote by $F[x]$ the set of all polynomials in the variable x with coefficients in F and if $f \in F[x]$, then we say that F is the **base field** of f. Looking back at some of the polynomials we encountered in Chap. 2, we see that

$$x^2 - 10x + 21 \in \mathbb{Q}[x], \quad y^3 - 15y - 4 \in \mathbb{Q}[y].$$

[1] The zero polynomial, in which f is identically zero, does not fit this definition and we do not assign it any notion of degree. Some authors set the degree of the zero polynomial to be $-\infty$ for technical reasons but this seems unnecessarily artificial.

[2] There are plenty of other types of field used in mathematics.

Equally, these polynomials could be considered as elements of $\mathbb{R}[x]$ and $\mathbb{R}[y]$, respectively, though $\mathbb{Q}$ is the smallest field that contains the coefficients of the polynomials. (The coefficients are in $\mathbb{Z}$ but $\mathbb{Z}$ is not a field.) Precisely because F is a field, we can add, subtract and multiply polynomials in $F[x]$ and get another polynomial in $F[x]$. It is a consequence of the division algorithm for polynomials [13, Thm. A.1.4] that if $f \in F[x]$ can be written as a product of polynomials $f = gh$, then both g and h will also be in $F[x]$.

The solutions to a polynomial equation of the form $f(x) = 0$ are referred to as the **roots** of the polynomial f and we will use this language (solutions of $f = 0$, or roots of f) interchangeably. An obvious question to ask when trying to find the roots of polynomials with degree higher than four is: are the complex numbers all that we need? They were introduced in order to make sense of Cardano's formula for cubics and then used in expressions for solutions to quadratics. Quartics didn't bring in the need for any new numbers, but if we start looking at polynomial equations of higher degree, will we need to add anything else into the mix? The answer is no. Complex numbers are all that we need and this is a consequence of a profound result, now known as the **fundamental theorem of algebra**.

This theorem implies that any polynomial of degree n, with arbitrary complex coefficients, can be factored into n linear factors:

$$a_n x^n + a_{n-1}x^{n-1} + \cdots + a_1 x + a_0 = a_n(x - x_1)(x - x_2)\cdots(x - x_n),$$

in which the numbers $x_1, x_2, \ldots, x_n$ are all complex numbers. Interestingly, while this is a theorem about algebra, its proof relies on non-algebraic concepts, such as the continuity of the real numbers, which are beyond the scope of this book. To give some idea of the flavour of the theorem, consider the cubic curves plotted in Fig. 2.2. They are all negative for large negative y and positive for large positive y and so we conclude (because the curves are continuous) that they must cross the $y = 0$ axis at least once. Hence a real cubic has at least one real root.

The fundamental theorem of algebra was stated in the seventeenth century, notably by Albert Girard in 1629, but the first attempts at a proof came in the eighteenth, with significant contributions from Jean le Rond d'Alembert. Many mathematicians, including the intellectual giants Leonhard Euler and Joseph-Louis Lagrange, tried to get to grips with the subtle ideas at play but the first to publish a rigorous proof was probably Jean-Robert Argand in 1806. Carl Friedrich Gauss, arguably the greatest mathematician of his time, published four different proofs, in 1799, 1815, 1816 and 1848, which perhaps gives a sense of the importance that was (and is) attached to the result. A comprehensive study of the fundamental theorem of algebra and its links to different branches of mathematics is given in [21]; see also [69, Chap. 8].

If the coefficient a_n of x^n is 1, a polynomial is said to be **monic** and for any given polynomial we can always divide through by a_n without affecting the value of the roots and so there is no loss of generality in considering only monic polynomials. The n roots $x_1, \ldots, x_n$ are not necessarily distinct. If the same root appears m times in this list then we say that the root has **multiplicity** m. Roots with multiplicity 1 are called **simple roots** whereas if $m > 1$ they are **multiple roots**. As an example, since

$$x^4 - 5x^3 + 9x^2 - 7x + 2 = (x - 1)^3(x - 2),$$

the roots of $x^4 - 5x^3 + 9x^2 - 7x + 2$ are 1 and 2, with 2 being a simple root and 1 a multiple root having multiplicity 3. In this way, counting roots according to their multiplicities, we see that a polynomial of degree n has precisely n roots, all in $\mathbb{C}$ thanks to the fundamental theorem of algebra.

With this general framework we can be more precise about our goal. Given a monic polynomial

$$f(x) = x^n + a_{n-1}x^{n-1} + \cdots + a_1x + a_0,$$

we wish to determine the n roots $x_1, \ldots, x_n$ in terms of the n coefficients $a_0, \ldots, a_{n-1}$. It is worth emphasising, as it lies at the heart of what is to come, that if $f \in F[x]$, where $F \subset \mathbb{C}$, then some, all, or none of the roots $x_1, \ldots, x_n$ may be in the field F, though the fundamental theorem of algebra tells us that they are all in $\mathbb{C}$. To give just one example, consider $x^2 - 2$, which is a polynomial in $\mathbb{Q}[x]$ and whose roots are $\pm\sqrt{2}$. But, as we noted at the beginning of this chapter, $\sqrt{2}$ cannot be written as a ratio of integers (it is **irrational**) and so neither of the roots of $x^2 - 2$ are in $\mathbb{Q}$.

At the end of Chap. 2, we made the remark that we needed a new way of looking at these problems and here is the first real glimpse of what that is going to entail. We start with a field F of coefficients and in some sense consider all numbers in F to be known quantities. Then we want to see in what way we need to extend F in such a way that we get a new field, let's call it K, such that the roots of the polynomial in question are all in K. Of course, the fundamental theorem of algebra tells us that we can choose $K = \mathbb{C}$ and we are done, but we will find that there are subfields of $\mathbb{C}$ that contain all the roots of any given polynomial and the nature of those fields will tell us a lot about the solutions.

3.2 Reducibility

The fundamental theorem of algebra shows that any polynomial can be written as a product of linear factors, each of the form $x + a$, where $a \in \mathbb{C}$. Some polynomials can be written as a product of factors where the coefficients in the factors belong to the base field. An example would be

$$x^2 - 10x + 21 = (x - 7)(x - 3)$$

since $x^2 - 10x + 21 \in \mathbb{Q}[x]$ and -3 and -7 are both elements of $\mathbb{Q}$. A polynomial $f \in F[x]$ is said to be **reducible** over F if it can be written as a product of two polynomials, both also in $F[x]$ and both of lower degree than f (or, equivalently, both of degree greater than zero). Thus $x^2 - 10x + 21$ is reducible over $\mathbb{Q}$.

Polynomials $f \in F[x]$ for which this is not possible are said to be **irreducible** over F. An example would be $x^2 + 1 \in \mathbb{Q}[x]$, since

$$x^2 + 1 = (x + \mathrm{i})(x - \mathrm{i})$$

but $\pm\mathrm{i}$ are not elements of $\mathbb{Q}$. The importance of the concept of irreducibility stems from the fact that any polynomial can be written as a product of irreducible factors, just as any integer can be written as a product of prime factors. This factorisation is unique, up to the order of the factors and multiplicative constants but may depend on the base field. The numbers $\pm\sqrt{2}$, which are the roots of $f(x) = x^2 - 2$, are not elements of $\mathbb{Q}$. Hence, considered as a polynomial over $\mathbb{Q}$, f is irreducible. But considered as a polynomial over $\mathbb{R}$ it is reducible since it can be factorised into $(x - \sqrt{2})(x + \sqrt{2})$. For brevity, we will sometimes write "$f \in F[x]$ is irreducible", or "f is irreducible over F", as shorthand for $f \in F[x]$ is irreducible over F. Note that the fundamental theorem of algebra is equivalent to the statement that all polynomials of degree at least 2 are reducible over the complex numbers.

If a polynomial has real coefficients then we can make a very important observation about the nature of its roots. First we note that if $z_1 = x_1 + \mathrm{i}y_1$ and $z_2 = x_2 + \mathrm{i}y_2$, then

$$z_1 z_2 = x_1 x_2 - y_1 y_2 + \mathrm{i}(x_1 y_2 + x_2 y_1)$$

and the product of the complex conjugates is

$$\overline{z_1}\,\overline{z_2} = x_1 x_2 - y_1 y_2 - \mathrm{i}(x_1 y_2 + x_2 y_1).$$

Hence the complex conjugate of $z_1 z_2$ is the product of $\overline{z_1}$ and $\overline{z_2}$. In particular, the complex conjugate of z^n is $(\overline{z})^n$. Since the complex conjugate of a real number is just the number itself, it follows that if α is a root of $f \in \mathbb{R}[x]$, in other words $f(\alpha) = 0$, then $f(\overline{\alpha}) = 0$ as well. Hence in this case the roots of f are either real or come in complex conjugate pairs (i.e. a pair $a + ib$ and $a - ib$). Therefore, if $f \in \mathbb{R}[x]$, it will have a number of factors of the form $(x - \beta)$, where β is real and a number of pairs of factors of the form $(x - \gamma)(x - \overline{\gamma})$, where γ is complex. Thus we have the factorisation

$$f = (x - \beta_1)\cdots(x - \beta_r)(x - \gamma_1)(x - \overline{\gamma}_1)\cdots(x - \gamma_s)(x - \overline{\gamma}_s),$$

where $\beta_1, \ldots, \beta_r$ are real and $r \geq 0$ is the number of real roots; $\gamma_1, \ldots, \gamma_s$ have non-zero imaginary part and $s \geq 0$ is the number of pairs of complex conjugate roots. Also $r + 2s = n$ since the total number of roots must be n. For any complex number z,

$$(x - z)(x - \overline{z}) = x^2 - 2x\,\mathrm{Re}\,z + |z|^2,$$

which is a quadratic polynomial in x with real coefficients and so any polynomial with real coefficients is reducible over $\mathbb{R}$ to a product of linear and quadratic factors.

For quadratic and cubic polynomials, reducibility over F is the same thing as the existence of roots in F, since at least one of the factors must be linear. However, for degrees greater than 3 this is not the case and it is possible that a polynomial of degree higher than 3 will be reducible over F but have no roots in F. The concept of irreducibility is incredibly important in what follows and so we will illustrate it with a number of examples. We will also begin to number our examples so that we can refer back to them later.

Example 3.1 $x^3 - 1 = (x - 1)(x^2 + x + 1)$ and so $x^3 - 1$ is reducible over $\mathbb{Q}$. Since $x^2 + x + 1 = (x - \omega)(x - \overline{\omega})$, where ω is the complex cube root of unity that we introduced in Chap. 2, this factor is irreducible over $\mathbb{Q}$ (and also over $\mathbb{R}$). ♦

Example 3.2 $x^4 - 5x^2 + 6 = (x^2 - 2)(x^2 - 3)$ is reducible over $\mathbb{Q}$ but none of its roots $(\pm\sqrt{2}, \pm\sqrt{3})$ is in $\mathbb{Q}$. ♦

Example 3.3 Let $f(x) = x^4 + a^4$. Since

$$f(x) = (x^2 + a\sqrt{2}x + a^2)(x^2 - a\sqrt{2}x + a^2)$$

we can use the quadratic formula to show that the roots of f are the four numbers

$$\tfrac{1}{2}a\sqrt{2}(1 \pm \mathrm{i}),\ \tfrac{1}{2}a\sqrt{2}(-1 \pm \mathrm{i}).$$

Hence if $a \in \mathbb{R}$, f is reducible over $\mathbb{R}$ but with no roots in $\mathbb{R}$. If $a = b\sqrt{2}$, where $b \in \mathbb{Q}$, then f is reducible over $\mathbb{Q}$ since then both $a\sqrt{2}$ and a^2 are rational. In particular, with $b = 1$,

$$x^4 + 4 = (x^2 + 2x + 2)(x^2 - 2x + 2).$$ ♦

A polynomial is called **separable** if it has no multiple roots, i.e. the roots $x_1, \ldots, x_n$ are all simple, otherwise it is **inseparable**. We won't have to worry too much about separability because we have the powerful result that:

Proposition 3.1 *If f is irreducible over any subfield of $\mathbb{C}$, then it is separable.*

This follows from [13, Lemma 5.3.5]. The essential idea is that $f \in F[x]$ being separable is equivalent to f and its derivative having no non-trivial common factors in $F[x]$. Irreducibility implies that this is the case. However, the converse of Proposition 3.1 is false. It is not the case that a polynomial that is separable over some subfield of $\mathbb{C}$ is necessarily irreducible. Irreducibility implies separability, but separability does not imply irreducibility.

Example 3.4 The polynomial $x^2 - 3x + 2 = (x-1)(x-2)$ is both separable and reducible over all subfields of $\mathbb{C}$. ♦

Irreducible polynomials are the basic building blocks used to construct all other polynomials. As we have already noted, any polynomial in $F[x]$ can be written, essentially uniquely, as a product of factors, each of which is irreducible over F. However, given a polynomial $f \in F[x]$, it is often a non-trivial task to determine whether it is irreducible. There are a number of approaches that we have at our disposal and we will describe a few here; a much more comprehensive set of tools is given in [55, Chap. 2]. First we have the **Rational Root Test**, which allows us to fairly easily determine if a polynomial with integer coefficients has any rational roots.

Rational Root Test
If $a_0, \ldots, a_n$ are integers and the fraction p/q (written in its lowest terms) is a rational root of $a_n x^n + a_{n-1}x^{n-1} + \cdots + a_1 x + a_0$, then p must divide the constant term a_0 and q must divide a_n.

To see this we note that if p/q is a root, then

$$a_n p^n + a_{n-1} p^{n-1} q + \cdots + a_1 p q^{n-1} + a_0 q^n = 0.$$

Clearly p is a factor of all but the last term and hence it must be a factor of that also. Similarly, q is a factor of all but the first term, and hence it must be a factor of that as well. Since p/q is assumed to be in its lowest terms, p and q have no common factors, and this is also true of any powers of p and q and so the result follows. In particular, if we have a monic polynomial ($a_n = 1$), with all the other coefficients integers, we can conclude that the only possible rational solutions are integers that divide the constant term a_0.

The rational root test provides an algorithm for determining whether or not a quadratic or cubic polynomial in $\mathbb{Q}[x]$ is irreducible over $\mathbb{Q}$, since the existence of roots in $\mathbb{Q}$ and reducibility over $\mathbb{Q}$ are equivalent for polynomials of degree 2 or 3. For a polynomial $f \in \mathbb{Q}[x]$ of degree higher than 3, the existence of a rational root certainly implies that f is reducible, but we cannot infer that f is irreducible over $\mathbb{Q}$ from the lack of a rational root.

Example 3.5 Any rational root of $x^3 + mx + 1$, where m is an integer, must divide 1 and hence must be ± 1. The number 1 is a root only if $m = -2$ and then

$$x^3 - 2x + 1 = (x - 1)(x^2 + x - 1).$$

The number -1 is a root only if $m = 0$ and then

$$x^3 + 1 = (x + 1)(x^2 - x + 1).$$

For all other integer values of m, $x^3 + mx + 1$ has no rational roots and hence is irreducible over $\mathbb{Q}$. ♦

Example 3.6 Consider

$$6x^5 - 5x^4 + 4x^3 - 4x^2 - 2x + 1.$$

Rational roots p/q must have a numerator p which divides 1 and a denominator q which divides 6. There are thus 6 possibilities: ± 1, $\pm 1/2$ and $\pm 1/3$. We can check them in turn and find that there are three rational roots: 1, 1/3 and $-1/2$. In fact we have the factorisation

$$6x^5 - 5x^4 + 4x^3 - 4x^2 - 2x + 1 = 6(x - 1)\left(x - \tfrac{1}{3}\right)\left(x + \tfrac{1}{2}\right)(x^2 + 1)$$

which shows that the other two roots are $\pm\, \mathrm{i}$. ♦

The second method that is available to us is **Gauss' lemma** which basically says that if we can show irreducibility over the integers, then irreducibility over the rationals comes for free. More precisely:

Gauss' Lemma
A non-constant polynomial in $\mathbb{Z}[x]$, whose coefficients have no common factors, is irreducible over $\mathbb{Z}$ if and only if it is irreducible over $\mathbb{Q}$.

This is article 42 of Gauss' *Disquisitiones Arithmeticae*, published in 1801; for a proof, see [13, Thm. A.3.2]. In particular this implies that if $g(x)$ and $h(x)$ are monic polynomials with rational coefficients and the product $g(x)h(x)$ is a polynomial with integer coefficients, then all the coefficients of $g(x)$ and $h(x)$ must be integers.

Example 3.7 Consider $f(x) = x^2 - 2$. By inspection this is clearly irreducible over the integers, so Gauss' lemma tells us that it is irreducible over the rationals as well, thus demonstrating the irrationality of $\sqrt{2}$.

Similarly, $x^2 - n$, where n is a non-square integer, is irreducible over $\mathbb{Z}$, and hence over $\mathbb{Q}$, and so $\sqrt{n}$ is irrational for any such n. ♦

If the base field is $\mathbb{Q}$, then we may be able to make use of **Eisenstein's irreducibility criterion**. This should perhaps more accurately be described as the Schönemann-Eisenstein criterion since Schönemann published it in 1846, four years before Eisenstein did. See [13, Thereom 4.2.3] for a proof.

Eisenstein's Irreducibility Criterion
If $a_0, a_1, \ldots, a_n$ are integers and p is prime, such that

- $a_0, a_1, \ldots, a_{n-1}$ are divisible by p, but a_n is not,
- a_0 is not divisible by p^2,

then $a_n x^n + a_{n-1} x^{n-1} + \cdots + a_1 x + a_0$ is irreducible over the rationals.

For a given polynomial in $\mathbb{Q}[x]$, we can multiply by the lowest common multiple of the denominators to yield a polynomial with integer coefficients and then see if we can find a prime p which satisfies the conditions.

Example 3.8 With p prime, $x^n + px + p$, $n \geq 2$ is irreducible over $\mathbb{Q}$. The same is true for $x^n + pax + p$, for any integer a. ♦

Example 3.9 $x^n - a \in \mathbb{Z}[x]$ is irreducible over $\mathbb{Q}$ for all $n \geq 1$, provided a is divisible by some prime p but not p^2. ♦

Example 3.10 $f = 2x^4 - \frac{3}{5}x^2 - \frac{6}{5}x + 3$ is irreducible over $\mathbb{Q}$ since $5f = 10x^4 - 3x^2 - 6x + 15$ is Eisenstein at 3 (meaning that it satisfies the Eisenstein criterion with $p = 3$). ♦

A trick that can sometimes be helpful is to note that if $f(x+a)$ is irreducible over F (where $a \in F$) then so is $f(x)$. This follows from the fact that any factorisation $f(x) = g(x)h(x)$ corresponds to a factorisation $f(x+a) = g(x+a)h(x+a)$.

Example 3.11 Let $f(x) = x^4 - 2x - 1$. Then $f(x+1) = x^4 + 4x^3 + 6x^2 + 2x - 2$. This is Eisenstein at 2. Hence f is irreducible over $\mathbb{Q}$. ♦

This device can be used to establish the important result that $x^{p-1} + \cdots + x + 1$ is irreducible over $\mathbb{Q}$ for any prime p:

Example 3.12 Consider

$$x^n - 1 = (x-1)(x^{n-1} + \cdots + x + 1).$$

If we write $x = y + 1$ and use the binomial theorem, then

$$\begin{aligned} \frac{x^n - 1}{x - 1} &= \frac{(y+1)^n - 1}{y} \\ &= y^{n-1} + \binom{n}{1} y^{n-2} + \binom{n}{2} y^{n-3} + \cdots + \binom{n}{n-2} y + n. \end{aligned}$$

(continued)

Example 3.12 (continued)
If n is prime, this satisfies the Eisenstein irreducibility criterion since each of the binomial coefficients is divisible by n and hence

$$x^{p-1} + \cdots + x + 1$$

is irreducible over $\mathbb{Q}$ for any prime p. ♦

Looking at all these examples of irreducilibty, we can observe that, in contrast to polynomials in $\mathbb{R}[x]$, where irreducible polynomials have degree 1 or 2, it is possible to have irreducible polynomials in $\mathbb{Q}[x]$ of arbitrarily large degree.

A given complex number α may or may not be the root of some polynomial in $F[x]$. If it is, we say that α is **algebraic** over F and if α is algebraic over $\mathbb{Q}$ then we call α an **algebraic number**. For example, there is no polynomial in $\mathbb{Q}[x]$ which has π as a root and so π is not an algebraic number (first proved by Ferdinand von Lindemann in 1882). On the other hand, $\sqrt{2}$ is an algebraic number since it is a root of $x^2 - 2$.

If α is the root of some polynomial $f \in F[x]$ then it will be the root of any polynomial fg, where $g \in F[x]$. In that case, we can define the **minimal polynomial** of α over F to be the unique monic polynomial $f \in F[x]$, of lowest degree, having α as a zero. It can be shown [13, Prop. 4.1.5] that

Proposition 3.2 *Let f be the minimal polynomial of α over F. Then*

- *f is irreducible over F (and hence separable; see Proposition 3.1),*
- *f is a factor of any other polynomial that has α as a root.*

Example 3.13 $x^2 - 2$ is the minimal polynomial of $\sqrt{2}$ over $\mathbb{Q}$. However, the minimal polynomial of $\sqrt{2}$ over $\mathbb{R}$ is $x - \sqrt{2}$. ♦

3.3 Roots of Unity

In Chap. 2, we showed how the concept of a cube root was linked to that of solving the equation $x^3 - 1 = 0$. Similarly, nth roots are linked to the solutions of $x^n - 1 = 0$, which are known as n**th roots of unity** and are fundamentally important in the solution of polynomial equations. Clearly 1 is a root for any n and since

$$x^n - 1 = (x - 1)(x^{n-1} + \cdots + x + 1),$$

the other $n - 1$ non-trivial roots of unity are also roots of $x^{n-1} + \cdots + x + 1$ which, as we showed in Example 3.12, is irreducible over $\mathbb{Q}$ if n is prime. Square roots of unity are ± 1 and the four fourth roots of unity are ± 1, $\pm\,\mathrm{i}$. We saw in Chap. 2 (see (2.9)) that the three cube roots of unity are $1, \omega$ and $\omega^2 = \overline{\omega}$, where

$$\omega = \tfrac{1}{2}(-1 + \mathrm{i}\sqrt{3}). \tag{3.1}$$

The only other case that we need to explore is that of the fifth roots of unity. These are 1, together with the four roots of the quartic

$$x^4 + x^3 + x^2 + x + 1,$$

which is irreducible since 5 is prime. We can solve this using Ferrari's method described in Chap. 2 and calculate that the four roots are

$$\tfrac{1}{4}(-1 + \sqrt{5}) \pm \tfrac{1}{4}\mathrm{i}\sqrt{10 + 2\sqrt{5}}, \quad \tfrac{1}{4}(-1 - \sqrt{5}) \pm \tfrac{1}{4}\mathrm{i}\sqrt{10 - 2\sqrt{5}}.$$

The root with positive real and imaginary parts will be denoted by ζ:

$$\zeta = \tfrac{1}{4}(-1 + \sqrt{5}) + \tfrac{1}{4}\mathrm{i}\sqrt{10 + 2\sqrt{5}}. \tag{3.2}$$

Since $(\zeta^m)^5 = (\zeta^5)^m = 1$ for any positive integer m, the other roots will be ζ^2, ζ^3 and ζ^4. Hence the five fifth roots of unity are $1, \zeta, \zeta^2, \zeta^3$ and ζ^4. Since $x^5 - 1$ is a polynomial with real coefficients, we know that its roots must either be real, or come in complex conjugate pairs. To that end, we note that $\zeta^3 = \overline{\zeta^2}$ and $\zeta^4 = \overline{\zeta}$, which is most easily established from the fact that if z is an nth root of unity it must have modulus 1 and hence $z\overline{z} = |z|^2 = 1$.

Throughout this book the symbols ω and ζ will always refer to the roots of unity given in (3.1) and (3.2). We noted in Chap. 2 that if $u^3 = a$ then the three cube roots of a are u, ωu and $\omega^2 u$. Similarly, if $u^5 = a$ then the five fifth roots of a are u, ζu, $\zeta^2 u$, $\zeta^3 u$ and $\zeta^4 u$.

To date we have thought of complex numbers purely algebraically, as numbers which satisfy the rules of arithmetic with the addition of the quantity $\mathrm{i} = \sqrt{-1}$. Another way of conceptualising them is to use a geometric representation that was first introduced by Caspar Wessel in 1797, but is named after Argand, who published the same idea in 1806. If we have the complex number with real part x and imaginary part y, i.e. $z = x + \mathrm{i}y$, then we can interpret this geometrically as the point (x, y) in what is now known as an Argand diagram. Figure 3.1 shows what we get if we plot the cube roots and fifth roots of unity using this device.

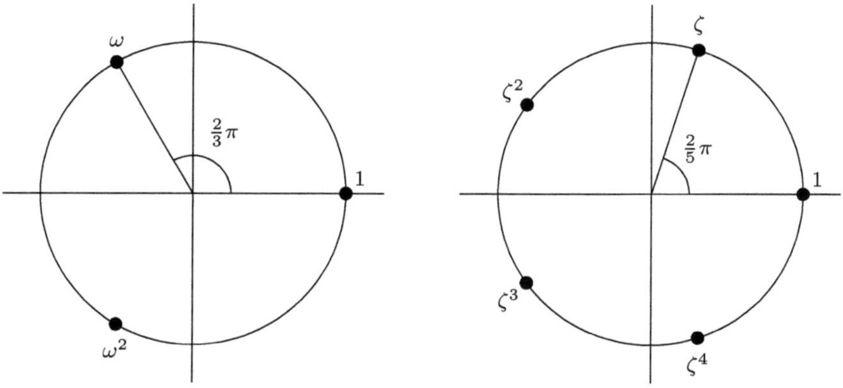

Fig. 3.1 Argand diagrams. Left: the cube roots of unity. Right: the fifth roots of unity

We see that $1, \omega, \omega^2$ are three points equally spaced around the unit circle, whereas $1, \zeta, \zeta^2, \zeta^3, \zeta^4$ are five such points.

Yet another way of representing complex numbers is in polar form. We saw in Chap. 2 that the modulus of z, denoted $|z|$, is the quantity $\sqrt{x^2+y^2}$. In an Argand diagram this is the distance of the number from the origin. Thus all the points on the unit circle represent complex numbers with a modulus of 1. We then define the **argument** of z, denoted $\arg z$, as the angle that the line connecting the number to the origin makes with the positive x-axis. Thus, from the figure, we can see that $\arg\omega = 2\pi/3$ and $\arg\zeta = 2\pi/5$. This leads to arguably the neatest representation of the roots of unity. We define the **complex exponential** $\mathrm{e}^{\mathrm{i}\theta}$ in terms of the standard trigonometric functions, via

$$\mathrm{e}^{\mathrm{i}\theta} = \cos\theta + \mathrm{i}\sin\theta,$$

often referred to as Euler's formula. Some elementary trigonometry then shows that

$$\omega = \mathrm{e}^{2\pi\mathrm{i}/3} \quad \text{and} \quad \zeta = \mathrm{e}^{2\pi\mathrm{i}/5}.$$

This is a device that we will use very infrequently as it involves non-algebraic functions, but sometimes solutions are written most compactly in this way and we include it for completeness.

It is clear from (3.1) and (3.2) that the cube and fifth roots of unity can be expressed using a finite number of the operations $+, -, \times, \div$ and the extraction of roots. In fact (see [3]), this is true for all roots of unity (i.e. all solutions to equations of the form $x^n - 1 = 0$, where n is a positive integer). This is equivalent to saying that the complex exponential $\mathrm{e}^{2\pi\mathrm{i}/n}$ can be written using a finite number of algebraic operations. Lagrange gave explicit formulas for $n \leq 10$, Vandermonde gave an incomplete proof in 1772 and Gauss later gave a complete proof (without reference to Vandermonde).

4 Symmetric Polynomials and Groups

Solving polynomial equations is naturally thought of as a process in which we construct the solutions in terms of the coefficients of the polynomial, whether they be specific numbers, as in $x^2 - 10x + 21$, or parameters, as in $x^2 + bx + c$. It was Lagrange who first appreciated that there was a lot to be gained by introducing symbols for the roots themselves and manipulating them as if they were known. Thus, for example, we have seen that the sum of the roots x_1 and x_2 of x^2+bx+c is $-b$ and their product is c. This idea allowed him to discover relationships between the roots of a polynomial without solving the polynomial explicitly and it proved pivotal in understanding when and how such equations can be solved. We now adopt this approach and begin by defining some particular combinations of the roots that underpin pretty much everything that follows.

4.1 Symmetric Polynomials

Thanks to the fundamental theorem of algebra we know that any monic polynomial of degree n can be written as

$$(x - x_1)(x - x_2) \cdots (x - x_n),$$

where $x_1, \ldots, x_n$ are complex numbers. What does this expression look like when we multiply it out? The coefficient of x^n is 1 as we need to take the x from each of the n factors. What about the coefficient of x^{n-1}? This will be made up from a sum of those terms where we form a product from $n - 1$ of the x's and one of the roots (and one minus sign); hence the coefficient of x^{n-1} is $-(x_1 + x_2 + \cdots + x_n)$. Similarly, the coefficient of x^{n-2} is made up from a sum of those terms in which we take $n - 2$ of the x's and two of the roots (and two minus signs); i.e. $x_1x_2 + x_1x_3 + \cdots + x_{n-1}x_n$.

C. M. Linton, *Quintic Equations and How to Solve Them*,
https://doi.org/10.1007/978-3-032-01658-4_4

This motivates the following definition of what are known as the **elementary symmetric polynomials** of the roots $x_1, \ldots, x_n$:

$$\begin{aligned} s_1^{(n)} &= x_1 + x_2 + \cdots + x_n, \\ s_2^{(n)} &= x_1x_2 + x_1x_3 + \cdots + x_{n-1}x_n, \\ &\vdots \\ s_n^{(n)} &= x_1x_2 \cdots x_n. \end{aligned}$$

Thus $s_1^{(n)}$ is the sum of the roots, $s_2^{(n)}$ is the sum of all the possible products of pairs of roots, and so on, with $s_n^{(n)}$ the sum of all possible products of n roots, of which there is only one. In terms of the elementary symmetric polynomials we have

$$(x - x_1)(x - x_2) \cdots (x - x_n) = x^n - s_1^{(n)}x^{n-1} + s_2^{(n)}x^{n-2} + \cdots + (-1)^n s_n^{(n)}.$$

We will come on to why these polynomials are called elementary and symmetric shortly, but if we compare this with the expression

$$x^n + a_{n-1}x^{n-1} + \cdots + a_1x + a_0,$$

we see that

$$s_1^{(n)} = -a_{n-1}, \quad s_2^{(n)} = a_{n-2}, \quad \ldots, \quad s_n^{(n)} = (-1)^n a_0. \tag{4.1}$$

Through this process we have actually discovered something quite profound about the roots of polynomials. If the coefficients $a_0, \ldots, a_{n-1}$ of a polynomial belong to a particular field F, then the elementary symmetric polynomials of the roots belong to the same field. For quadratic equations we have $s_1^{(2)} = x_1 + x_2$ and $s_2^{(2)} = x_1x_2$ and thus if the coefficients of our quadratic are rational, then the sum and product of the roots will be rational even though the roots themselves may be irrational. Similarly, for a cubic equation with rational coefficients, the quantities $x_1 + x_2 + x_3$, $x_1x_2 + x_1x_3 + x_2x_3$ and $x_1x_2x_3$ are all rational. And so on for higher-order equations.

Example 4.1 $x^2 - 2 \in \mathbb{Q}[x]$ has roots $\pm\sqrt{2}$, neither of which are rational. However, their sum (zero) and their product ($-$ 2) are. ♦

Example 4.2 The three cube roots of unity $1, \omega, \omega^2$ are roots of $x^2 - 1 \in \mathbb{Q}[x]$, with ω a root of $x^2 + x + 1$. The sum $1 + \omega + \omega^2$ is therefore zero and the sum of the pairwise products is $\omega + \omega^2 + \omega^3 = \omega + \omega^2 + 1$ which is again zero. Finally, their product is $\omega^3 = 1$. ♦

We started Chap. 2 with a problem where the sum and product of two numbers were given; in other words, we specified $s_1^{(2)}$ and $s_2^{(2)}$ and sought to determine x_1 and x_2. In fact, finding the roots of $x^n + a_{n-1}x^{n-1} + \cdots + a_1 x + a_0$ is equivalent to solving the set of equations in (4.1) for $x_1, \ldots, x_n$ given $s_1^{(n)}, \ldots, s_n^{(n)}$. For example, finding the roots to the cubic equation $x^3 + px + q = 0$ is precisely the same problem as determining three numbers x_1, x_2, x_3 which satisfy

$$x_1 + x_2 + x_3 = 0, \quad x_1x_2 + x_1x_3 + x_2x_3 = p, \quad x_1x_2x_3 = -q.$$

In general, an nth degree equation can be thought of as a single condition satisfied by each of its n roots $x_1, \ldots, x_n$. Equally, those roots can also be thought of as a solution to the system of n equations given by (4.1) in which each root appears in exactly the same way; we might say that they appear symmetrically.

To be a bit more precise, let's pose the question: in what sense are $s_1^{(n)}, \ldots, s_n^{(n)}$ symmetric? To answer this we will delve into the world of permutations. If we consider the set $X = \{x_1, \ldots, x_n\}$ then we can define a **permutation** σ acting on X as a function which has the effect of reordering the elements of X.[1]

Example 4.3 Using the two line notation introduced by Augustin-Louis Cauchy, the permutation

$$\sigma = \begin{pmatrix} x_1 & x_2 & x_3 & x_4 & x_5 & x_6 \\ x_2 & x_3 & x_1 & x_4 & x_6 & x_5 \end{pmatrix}$$

sends x_1 to x_2, x_2 to x_3, x_3 to x_1, leaves x_4 fixed, and interchanges x_5 and x_6. Alternatively, we can write

$$\sigma\{x_1, x_2, x_3, x_4, x_5, x_6\} = \{x_2, x_3, x_1, x_4, x_6, x_5\}. \quad \ldots$$

[1] Historically, the word permutation was used to denote a particular arrangement of the elements of X, with the word substitution used for the function which rearranges the elements.

In the case of a **cyclic permutation**, where the elements are simply shifted round in a cyclic fashion, we can just use a single line. Thus $(x_1x_2x_3)$ is the permutation that takes x_1 to x_2, x_2 to x_3 and x_3 to x_1. Clearly $(x_1x_2x_3) = (x_3x_1x_2) = (x_2x_3x_1)$, but $(x_1x_3x_2)$ is different. We can extend the applicability of a permutation σ by allowing it to act on some function of the quantities $x_1, \dots, x_n$, which in our case will be polynomials. If we have a function $f(x_1, \dots, x_n)$ then the action of a permutation σ on f is defined in the obvious way, i.e. $\sigma f(x_1, \dots, x_n) = f(\sigma(x_1, \dots, x_n))$. For example, if $f = x_1x_2 + x_3$ and $\sigma = (x_1x_2x_3)$, then $\sigma f = x_2x_3 + x_1$.

Not all permutations are cyclic, but it is possible to write any permutation as a combination of disjoint cycles, i.e. cyclic permutations in which each element appears once and only once. To obtain this representation, start with x_1 and follow it through successive permutations until you return to x_1, then move to the next element of the set excluding the ones in the first cycle, and so on.

...Continuing Example 4.3, we find through this process that

$$\sigma = (x_1x_2x_3)(x_4)(x_5x_6).$$

The (x_4) term, which says that x_4 is unchanged by the permutation, can be omitted though it is often useful to retain it as it makes explicit the underlying set on which the permutation is acting. ♦

For a given permutation, we can define the **cycle type** to be a list containing the lengths of the disjoint cycles that make up σ, in decreasing order.

Example 4.4 The cycle type of the permutation in Example 4.3 is (3, 2, 1). The cycle type of $(x_1x_5)(x_2)(x_3)(x_4x_6)$ is (2, 2, 1, 1). ♦

A key observation about the elementary symmetric polynomials is that they remain unchanged by any permutation of $\{x_1, \dots, x_n\}$; for example, if we apply any such permutation to $x_1 + \cdots + x_n$ it will still be $x_1 + \cdots + x_n$ because the order in which we do the additions is immaterial. Another way of saying the same thing is that it doesn't matter which of the roots of our polynomial that we label as x_1, which as x_2 and so on; the elementary symmetric polynomials of the roots will remain unaltered. Of course, there are many more polynomials with this property than just $s_1^{(n)}, \dots, s_n^{(n)}$, for example $x_1^2 + \cdots + x_n^2$. Any polynomial in the variables $x_1, \dots, x_n$ which is unchanged by any permutation of $\{x_1, \dots, x_n\}$ is called a **symmetric polynomial**.

We then have the very important result, known as the fundamental theorem of symmetric polynomials [13, Thm. 2.2.2]:

The Fundamental Theorem of Symmetric Polynomials
Any symmetric polynomial in the variables $x_1, \ldots, x_n$ (with coefficients in a field F) is a polynomial in $s_1^{(n)}, \ldots, s_n^{(n)}$ (with coefficients in F).

It is in this sense that $s_1^{(n)}, \ldots, s_n^{(n)}$ are elementary; all other symmetric polynomials can be expressed in terms of them. This theorem appears to have been understood by Isaac Newton though first formulated precisely as a theorem by Lagrange. The fundamental theorem of symmetric polynomials provides us with a great deal of information about the roots of a polynomial. Since the elementary symmetric polynomials of the roots are, up to factor of ± 1, equal to the coefficients of the polynomial (as in (4.1)), it tells us that *any* symmetric polynomial in the roots is a polynomial in the coefficients. As a consequence, if the polynomial has coefficients in a field F, then any symmetric function of the roots is in F. This is a significant result. Without knowing the solutions, we have deduced a property of some combinations of the roots of a polynomial. Moreover, the characteristic that these combinations of the roots share is defined in terms of the effect caused by permuting the roots. This will be a recurring theme.

The degree of the polynomial under discussion is usually clear from the context and then we can dispense with the superscript (n) on the symmetric polynomials, but it should always be borne in mind that the quantities $s_1, s_2, \ldots$ depend on n.

Example 4.5 For $n \geq 2$, by expanding out the product s_1^2 it can be shown that

$$x_1^2 + x_2^2 + \cdots + x_n^2 = s_1^2 - 2s_2.$$

For $n = 2$, if we take the polynomial $x^2 + bx + c$, then the sum of the roots is $-b$ and the product is c. Then we have

$$x_1^2 + x_2^2 = b^2 - 2c. \quad \blacklozenge$$

Example 4.6 For $n = 3$ we can establish, with a bit of effort,

$$x_1^2x_2^2 + x_1^2x_3^2 + x_2^2x_3^2 = s_2^2 - 2s_1s_3$$

and

$$x_1^3 + x_2^3 + x_3^3 = s_1^3 - 3s_1s_2 + 3s_3.$$

Hence the three roots of the cubic $x^3 + px + q$ satisfy

$$x_1^2x_2^2 + x_1^2x_3^2 + x_2^2x_3^2 = p^2, \quad x_1^3 + x_2^3 + x_3^3 = -3q.$$ ♦

4.2 Permutations and Groups

Thought of as functions, the composition of two permutations, $\tau \circ \sigma$ is the result of first applying σ and then applying τ (i.e $(\tau \circ \sigma)(x) = \tau(\sigma(x))$). Thought of as permutations, it is more natural to write the product of applying σ first and then τ as $\sigma\tau$ and this is the convention we follow here. Thus $\sigma\tau$ is equivalent to $\tau \circ \sigma$. We observe a number of things about permutations:

1. If σ and τ are two permutations then $\sigma\tau$ is also a permutation.
2. Since function composition is always associative, i.e. $f \circ (g \circ h) = (f \circ g) \circ h$, so it is for products of permutations: $\sigma(\tau\mu) = (\sigma\tau)\mu$.
3. There is a permutation that leaves the order of the elements unchanged, we call this the identity permutation and denote it by *id*.
4. For any permutation σ acting on $\{x_1, \ldots, x_n\}$, there is another permutation τ which reverts the set back to $\{x_1, \ldots, x_n\}$, i.e. $\sigma\tau = \tau\sigma = id$.

These properties show that the set of permutations of $\{x_1, \ldots, x_n\}$ form a mathematical object known as a **group** and this specific group is known as the **symmetric group**, denoted by S_n. The **order** of a group is the number of elements in that group and the order of S_n, written $|S_n|$, is the number of ways of arranging n objects. The first in the list can be any one of the n objects, the second is then one of $n-1$ objects and so on; hence

$$|S_n| = n(n-1)\cdots 2 \times 1 = n!$$

While we have introduced the concept of a group in terms of permutations, it is important to recognise that groups come in many different forms, as we shall see. Any set of elements that can be combined according to the four properties above, which we call, respectively, the **closure**, **associativity**, **identity** and **inverse properties** is a group. There are traces of the ideas that now form the basis of group theory in the work of a number of eighteenth century mathematicians, notably Lagrange, but it was the work of Galois and Cauchy in the mid-nineteenth century that started to put things on firm foundations. The word group (*groupe* in French) was coined by Galois.

In terms of this new language, a symmetric polynomial is one which remains unaltered by the action of any element of the symmetric group S_n. Note that, in general, permutations do not commute, by which we mean that it is not true in general that $\sigma\tau = \tau\sigma$. For example,

$$(x_1x_2)(x_1x_3) = (x_1x_2x_3), \quad (x_1x_3)(x_1x_2) = (x_1x_3x_2).$$

(Remember that the permutation on the left comes first and so in the first of these examples x_1 becomes x_2 from the first permutation and then is unaffected by the second, and so on.) If we do have $\sigma\tau = \tau\sigma$ for any two elements in the group then the group is said to be **abelian**, named after Niels Henrik Abel who, as we shall see, played a crucial role in the theory of quintic equations.

A permutation which simply interchanges two elements is called a **transposition** and any permutation can be written as a product of transpositions. This follows since any cycle can be written in this way. Thus, for example, an n-cycle can be written as a product of $n-1$ transpositions:

$$(x_1x_2x_3\cdots x_n) = (x_1x_2)(x_1x_3)\cdots(x_1x_n). \tag{4.2}$$

This representation in terms of transpositions is not unique though. For example, we also have

$$(x_1x_2x_3\cdots x_n) = (x_{n-1}x_n)(x_{n-2}x_{n-1})\cdots(x_1x_2),$$

but it is not difficult to show that the number of transpositions in any representation will always be either even or odd. You can't write the product of an even number of transpositions as a product of an odd number of them, or vice versa. Thus we can define an **even** permutation to be one that can be represented by an even number of transpositions and likewise for an **odd** permutation.

It follows from (4.2) that a cycle of length n will be an even permutation if n is odd and an odd permutation if n is even. If a permutation σ_1 can be represented by r_1 transpositions and σ_2 by r_2 transpositions, then the product $\sigma_1\sigma_2$ can be represented by $r_1 + r_2$ transpositions. Hence we see that the product of two even permutations or two odd permutations is even, whereas if we compose an even and an odd permutation, in either order, an odd permutation results. Also the inverse of a permutation has the same parity as the permutation itself, since a transposition is its own inverse and so inverting a product of transpositions reverses the order but doesn't change the number of transpositions. This all leads us to the important observation that the even permutations form a **subgroup** of S_n (a subset of S_n that is itself a group) known as the **alternating group**, denoted by A_n, and $|A_n| = n!/2$. (The odd permutations do not form a subgroup since the product of two odd permutations is not odd.)

4.3 The Discriminant

We are now in a position to examine arguably the most important symmetric function of the roots of a polynomial, namely its **discriminant** Δ, which is defined as the product of the squares of the differences between the roots. For a polynomial with roots $x_1, \dots, x_n$ we have

$$\Delta = (x_1 - x_2)^2(x_1 - x_3)^2 \cdots (x_{n-1} - x_n)^2.$$

Clearly the discriminant is non-zero if and only if all the roots are distinct. In other words, if $\Delta \neq 0$, the polynomial is separable, whereas if $\Delta = 0$, it has at least one multiple root. Since Δ is a symmetric function of the roots, it can be expressed as a polynomial function of the coefficients of the original polynomial. Hence, if $f \in F[x]$, we have $\Delta \in F$. It follows from the definition of the discriminant that

Proposition 4.1 *If all the roots of a polynomial are real, then* $\Delta \geq 0$.

We also define

$$\delta = (x_1 - x_2)(x_1 - x_3) \cdots (x_{n-1} - x_n),$$

so that $\delta^2 = \Delta$. Unlike the discriminant, δ is not a symmetric polynomial since, for example, the permutation which swaps x_1 and x_2 changes the factor $(x_1 - x_2)$ to $(x_2 - x_1)$ and leaves the rest of the product as before. For a given permutation $\sigma \in S_n$, then either $\sigma\delta = \delta$, which will happen if σ is an even permutation, or $\sigma\delta = -\delta$, which will happen if σ is an odd permutation. Hence δ is invariant under the action of any permutation in the alternating group, A_n, but not for other permutations in S_n.

The discriminant plays a key role in the solutions for quadratic and cubic equations that we have described previously. For the quadratic equation $x^2 + bx + c$, we have

$$\Delta = (x_1 - x_2)^2 = (x_1 + x_2)^2 - 4x_1x_2 = b^2 - 4c.$$

In this equation the first equality is just the definition of the discriminant for a polynomial of degree 2, the second equality is the representation of a symmetric polynomial in terms of elementary symmetric polynomials and the final equality uses the expressions for those elementary symmetric polynomials in terms of the coefficients of the original equation. The two roots of the quadratic, when written in terms of the discriminant, are thus

$$\tfrac{1}{2}(-b \pm \sqrt{\Delta}).$$

For cubic equations things are somewhat more complicated. However, we can begin by noting that in Chap. 2 we transformed the cubic $x^3 + bx^2 + cx + d$ into the depressed cubic

$$y^3 + py + q$$

(with $p = c - b^2/3$, $q = d + 2b^3/27 - bc/3$) via the transformation $x = y - b/3$. Since this is just a translation, the differences between the roots x_1, x_2, x_3 of the original cubic are exactly the same as the differences between the roots y_1, y_2, y_3 of the depressed cubic; thus the discriminants of the two polynomials are the same. The elementary symmetric polynomials for the depressed cubic are

$$y_1 + y_2 + y_3 = 0, \qquad y_1y_2 + y_1y_3 + y_2y_3 = p, \qquad y_1y_2y_3 = -q$$

and with the aid of these formulas, and the formulas given in Examples 4.5 and 4.6, we can show that

$$\Delta = (y_1 - y_2)^2(y_1 - y_3)^2(y_2 - y_3)^2 = -4p^3 - 27q^2.$$

It would be misleading to say that this calculation is straightforward, however. There are some clever ways of deriving formulas for discriminants that reduce the complexity of the calculations, but they involve ideas that are beyond our scope. Looking back at Cardano's formula, we see that it involves taking the square root of a quantity $D = -\Delta/108$. In fact we have

$$y = \sqrt[3]{-\tfrac{1}{2}q + \tfrac{1}{18}\sqrt{-3\Delta}} + \sqrt[3]{-\tfrac{1}{2}q - \tfrac{1}{18}\sqrt{-3\Delta}}$$

and once again we see that the discriminant plays a prominent role in the solution. The term 'discriminant' was coined by James Joseph Sylvester[2] in the mid-nineteenth century because the sign of Δ allows us to discriminate between different types of solution.

When p and q are real, we know that either the roots are all real or there is a single real root with the other two forming a complex conjugate pair (see Sect. 3.2). In this situation, if we work out the discriminant for the case of one real root and

[2] Note that Joseph is not a second given name, but rather Sylvester is a second family name that was adopted when James was at univesity.

two complex conjugate roots, we find that $\Delta < 0$, whereas if all the roots are real then $\Delta \geq 0$. If $\Delta < 0$, then the numbers under the cube roots in Cardano's formula are real and taking the real cube roots in each case leads to a real root $u + v$, while the other two roots $\omega^2 u + \omega v$ and $\omega u + \omega^2 v$ are complex conjugates. If $\Delta = 0$, then at least two of the roots are equal (and hence real) and so all the roots must be real. Finally, if $\Delta > 0$, then Cardano's formula requires that we take the square root of a negative quantity even though the three roots must be real. In the sixteenth century, due to this counterintuitive result, this became known as the *Casus Irreducibilis*. Note that this does not imply that the cubic is irreducible in the sense of Sect. 3.2.

Calculating the discriminant of a quartic direct from the definition is a daunting task and the results are long, complicated, and hardly illuminating. A much more significant observation is that the discriminant of a quartic is the same as the discriminant of the cubic equation that forms part of its solution, whether one uses Ferrari's method or that of Descartes. We will demonstrate this for Ferrari's approach, where the cubic is given in (2.13). Referring back to the solution from Chap. 2 we see that the four solutions to the quartic x_1, x_2, x_3, x_4 are the solutions to two quadratic equations (given in (2.15)) and therefore that two of the roots x_1 and x_2 satisfy

$$x_1 x_2 = \tfrac{1}{2} y_1 + D,$$

where y_1 is one of the three roots to the cubic (2.13) and the other two, x_3 and x_4, satisfy

$$x_3 x_4 = \tfrac{1}{2} y_1 - D,$$

from which

$$x_1 x_2 + x_3 x_4 = y_1.$$

The cubic has three roots and it doesn't matter which we choose. Different choices result in different pair combinations in the above. Thus if $y = y_2$, say, x_1 and x_3 are paired and for $y = y_3$, x_1 and x_4 go together. Thus we have

$$x_1 x_3 + x_2 x_4 = y_2, \qquad x_1 x_4 + x_2 x_3 = y_3.$$

It follows that

$$\begin{aligned} y_1 - y_2 &= (x_1 - x_4)(x_2 - x_3), \\ y_1 - y_3 &= (x_1 - x_3)(x_2 - x_4), \\ y_2 - y_3 &= (x_1 - x_2)(x_3 - x_4). \end{aligned}$$

Hence the discriminant of the cubic (which is the product of the squares of the left-hand sides of these equations) is identical to the discriminant of the original quartic (which is the product of the squares of the right-hand sides of these equations). A similar analysis can be carried out for the cubic (2.18) in Descartes method, with the same conclusion: its discriminant is identical to that of the quartic from which it is derived.

4.4 The Symmetric Groups S_2, S_3 and S_4

We will return to our discussion of the roots of polynomials shortly, but first we need to look in further detail at permutation groups and, in particular, examine the symmetric groups S_2, S_3 and S_4 as these groups are crucial to the understanding of the structure of the roots of quadratics, cubics and quartics. In the process we will introduce a number of important concepts from group theory, including cyclic groups, isomorphism, subgroups, normal subgroups, quotient groups, conjugacy classes and simple groups.

We begin with S_2 which contains all the permutations of the set $\{x_1, x_2\}$. There are just two: the identity permutation, *id*, and the permutation $\sigma = (x_1x_2)$ which interchanges the two elements. Obviously, performing σ twice gets us back to the original order and so we see that

$$S_2 = \{id, \sigma\}, \quad \sigma^2 = id.$$

Another way of looking at S_2 is to say that it is **generated** by σ. In other words, all the elements of S_2 can be expressed in terms of powers of σ. We write this as

$$S_2 = \langle\sigma\rangle, \quad \sigma^2 = id.$$

There is little to be gained by this representation in this case, but for larger groups it will be convenient not to have to list out all the elements. As we will see, most groups are not generated by a single element. For example, we write $\langle\sigma, \tau\rangle$ for the group that consists of all the elements that can be written as a product of some power (positive, negative or zero) of σ and some power of τ. Note that σ^{-1} (the inverse of σ) is the element which satisfies $\sigma\sigma^{-1} = \sigma^{-1}\sigma = id$.

Groups which are generated by a single element, i.e. $\langle\sigma\rangle$, are called **cyclic groups** and the order of a cyclic group is the smallest integer n such that $\sigma^n = id$. The elements of the group are $id, \sigma, \ldots, \sigma^{n-1}$.

Example 4.7 Let $\sigma = (x_1x_2x_3)$, a cyclic permutation of order 3. Then $\sigma^2 = (x_1x_3x_2)$ and $\sigma^3 = id$. The group generated by σ is therefore $\{id, (x_1x_2x_3), (x_1x_3x_2)\}$ and hence has order 3. ♦

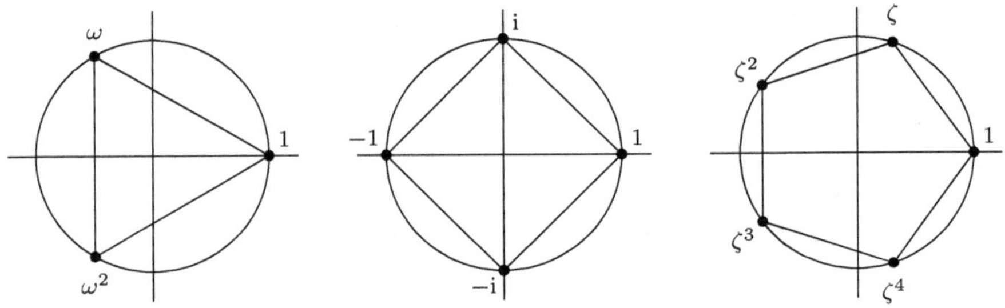

Fig. 4.1 From left to right: cube roots of unity, fourth roots of unity, fifth roots of unity

The concept of a group is much more general than just the permutation groups that we have considered so far and we have actually already encountered other types of group, though we did not identify them as such. Thus the cube roots of unity are $\{1, \omega, \omega^2\}$, with $\omega^3 = 1$ and the fifth roots of unity are $\{1, \zeta, \zeta^2, \zeta^3, \zeta^4\}$, with $\zeta^5 = 1$. These are cyclic groups of orders 3 and 5, respectively, though now the group operation is multiplication rather than function composition.

Another way of looking at these cyclic groups is to use the geometric representation of complex numbers that we described in Sect. 3.3. Figure 4.1 shows what we get if we plot the powers of ω and ζ using this device and consider them as vertices of a regular polygon. We see that $1, \omega, \omega^2$ lie at the vertices of an equilateral triangle located on the circle $|z| = 1$, whereas $1, \zeta, \zeta^2, \zeta^3, \zeta^4$ form a regular pentagon. The fourth roots of unity, ± 1, $\pm \mathrm{i}$ can be seen to mark out a square.

The diagrams in the figure show that we can also consider cyclic groups as rotations. Rotations through $360°/3 = 120°$ form a cyclic group of order 3 and are the rotational symmetries of an equilateral triangle. Rotations through $360°/4 = 90°$ form a cyclic group of order 4 and are the rotational symmetries of a square. Similarly, rotations through $360°/5 = 72°$ form a cyclic group of order 5 and are the rotational symmetries of a regular pentagon.

We thus have lots of examples of cyclic groups, but their structure only depends on the number of elements in the group. The rotational symmetries of an equilateral triangle, the complex numbers $1, \omega, \omega^2$ under multiplication, and the permutation group generated by $(x_1x_2x_3)$ are all, in some abstract sense, the same; they can all be written as

$$\langle\sigma\rangle, \quad \sigma^3 = id,$$

for some group element σ. This is a very powerful idea; it allows us to describe properties of a class of groups all at the same time and the results will be applicable to any particular manifestation of that class of group. A cyclic group of order n is denoted by C_n and we say that any group G which has the structure of C_n is **isomorphic** to C_n, written $G \cong C_n$. Every cyclic group is abelian since any two elements σ^k, σ^m satisfy $\sigma^k\sigma^m = \sigma^{k+m} = \sigma^{m+k} = \sigma^m\sigma^k$. Going right back to the start of this discussion we see that

$$S_2 \cong C_2.$$

We turn now to S_3, the possible permutations of the set $\{x_1, x_2, x_3\}$, which has $3! = 6$ elements. They are

$$id,\ (x_1x_2x_3),\ (x_1x_3x_2),\ (x_2x_3),\ (x_1x_3),\ (x_1x_2);$$

the identity, two 3-cycles, and three transpositions which leave the third element fixed. If we set $\sigma = (x_1x_2x_3)$ and $\tau = (x_2x_3)$ (we could have chosen either of the 3-cycles and any one of the transpositions), then we can show that[3]

$$S_3 = \langle\sigma, \tau\rangle = \{id, \sigma, \sigma^2, \tau, \sigma\tau, \sigma^2\tau\},$$

with $\sigma^3 = \tau^2 = id,\ \tau\sigma = \sigma^2\tau$.

Unlike S_2, the group S_3 is not abelian; its elements do not in general commute. For example

$$\sigma\tau = (x_1x_2x_3)(x_2x_3) = (x_1x_3),$$
$$\tau\sigma = (x_2x_3)(x_1x_2x_3) = (x_1x_2).$$

In fact S_3 is the smallest non-abelian group. An important property that we will need to understand is whether S_3 has any subgroups or not and if so, what they look like. A subgroup G of S_3, written $G \leq S_3$, will be some subset of $\{id, \sigma, \sigma^2, \tau, \sigma\tau, \sigma^2\tau\}$ that is itself a group. Technically S_3 is a subgroup of itself, and $\{id\}$, the trivial group, is a subgroup of any group. Any subgroup other than the whole group or the trivial group is known as a **proper subgroup** and it is these that we seek to identify. (Note that we use the term proper subgroup slightly differently to the terms proper subset and proper subfield in that we rule out the trivial group as well as the group itself.)

In fact we have already identified one proper subgroup above, in Example 4.7, namely $\{id, (x_1x_2x_3), (x_1x_3x_2)\}$ which is just $\{id, \sigma, \sigma^2\}$ with $\sigma^3 = id$. The non-trivial elements of this group are 3-cycles and thus even permutations, which shows that this group is the alternating group A_3 and we have

$$A_3 \cong C_3.$$

[3] This is a special case of the fact that the symmetric group S_n is generated by the n-cycle $(x_1x_2 \ldots x_n)$ and any single transposition (x_jx_k) for which $|j - k|$ and n are co-prime. https://kconrad.math.uconn.edu/blurbs/grouptheory/genset.pdf, Thm. 2.8.

There are also three subgroups of order 2: $\{id, \tau\}$, $\{id, \sigma\tau\}$, $\{id, \sigma^2\tau\}$ each of which consists of the identity plus a transposition and is isomorphic to C_2. A famous theorem in group theory, known as Lagrange's theorem, states that:

Lagrange's Theorem
The order of any subgroup divides the order of the group itself, the ratio being known as the **index** of the subgroup.

It perhaps should not pass without comment that a theorem in group theory is named after someone writing long before there was such a thing as group theory! Lagrange's actual statement is described later in Sect. 7.1. But group theory did not just appear overnight, fully formed, it emerged over decades as the amalgamation and synthesis of a great many ideas that were floating around at the time. Although (a version of) this theorem was stated by Lagrange it wasn't proved for a general permutation group until the work of Camille Jordan in 1861. Thus for S_3, proper subgroups must have order 2 or 3, as we have observed. The index of A_3 in S_3 is $6/3 = 2$ and the index of any of the three subgroups of order 2 is $6/2 = 3$.

Some subgroups have a particularly nice property that allows us to define the group equivalent of division and this will be crucial when we start to link group structures to the process of solving polynomial equations. Before we can describe such subgroups we need to introduce the concept of a coset. If we have a subgroup H of a group G then we can use it to partition G into disjoint sets called **cosets**. Due to the non-abelian nature of groups in general we can do this in two ways, using left cosets or right cosets. The left coset associated with the element $g \in G$ and the subgroup H is defined as

$$gH = \{gh, h \in H\},$$

i.e. all the elements of G that are the products of g (on the left) with an element of H. Similarly the right coset associated with g is $Hg = \{hg, h \in H\}$. Obviously, if G is abelian these are the same thing. The subgroup H itself is both a left and right coset since we can take $g = id$. Every left or right coset of H has the same number of elements as H and, by Lagrange's theorem, $|H|$ divides $|G|$.

Example 4.8 Let's work out the cosets of some elements of $G = S_3$ using the subgroup $H = A_3$:

$$id\,A_3 = \{id, \sigma, \sigma^2\}, \quad \tau A_3 = \{\tau, \tau\sigma, \tau\sigma^2\} = \{\tau, \sigma^2\tau, \sigma\tau\},$$

$$A_3\,id = \{id, \sigma, \sigma^2\}, \quad A_3\tau = \{\tau, \sigma\tau, \sigma^2\tau\},$$

(continued)

Example 4.8 (continued)

where we have used $\sigma^3 = \tau^2 = id$, $\tau\sigma = \sigma^2\tau$ to express all products of σ and τ in the form $\sigma^j\tau^k$. The top line lists the left cosets of *id* and τ and these partition the six elements of S_3 into two disjoint sets. Note that the left cosets generated by *id*, σ and σ^2 are the same and similarly, $\tau A_3 = \tau\sigma A_3 = \tau\sigma^2 A_3$. The lower line does the same for right cosets and the two left cosets are seen to be identical to the two right cosets. ♦

It is fairly straightforward to show that, as seen in this example, the number of cosets associated with a subgroup H of G is equal to $|G|/|H|$. However, it is not always true that the left and right cosets are the same.

Example 4.9 Take $G = S_3$ again, but now let $H = \{id, \tau\}$. We can calculate

$$id\, H = H\, id = \{id, \tau\},$$

$$\sigma H = \{\sigma, \sigma\tau\}, \quad \sigma^2 H = \{\sigma^2, \sigma^2\tau\},$$

$$H\sigma = \{\sigma, \tau\sigma\} = \{\sigma, \sigma^2\tau\}, \quad H\sigma^2 = \{\sigma^2, \tau\sigma^2\} = \{\sigma^2, \sigma\tau\}.$$

We thus see that for this choice of H, the set of left cosets is not the same as the set of right cosets though there are the same number of each, again given by $|G|/|H|$. ♦

The introduction of the cosets of a group G associated with a subgroup H allows us to create a set with $|G|/|H|$ elements. What we would like is for that set to itself be a group as then we can in some sense divide one group by another with a group as the result. A group is a set of elements together with an operation that combines any two elements (satisfying certain properties) and so this will only make sense if we can define an appropriate operation between cosets. We will use the symbol $*$ to represent that operation and then, if we consider the case of left cosets, the natural way to define $*$ is via

$$(\sigma_1 H) * (\sigma_2 H) = \sigma_1\sigma_2 H,$$

with a similar definition for right cosets. However, as we saw in the preceding examples, we can have two (or more) different elements of G, let's call them σ_1 and σ_1', such that the cosets $\sigma_1 H$ and $\sigma_1' H$ are identical. And, similarly, we can have two different elements σ_2 and σ_2', such that $\sigma_2 H = \sigma_2' H$. So, in order for our definition of $*$ to make sense it must be the case that in these circumstances $\sigma_1\sigma_2 H = \sigma_1'\sigma_2' H$.

It turns out that this is equivalent to insisting that the left and right cosets of H are the same, as in the case of A_3 in S_3 above.

If the left and right cosets of H are the same, then we say that H is a **normal subgroup** of G, written $H \triangleleft G$ (or $G \triangleright H$). If G is an abelian group, the left and right cosets of any subgroup H are identical and hence:

Proposition 4.2 *Any subgroup of an abelian group is a normal subgroup.*

The set of cosets, with the group operation $*$ as defined above, can then be shown to form a group with order $|G|/|H|$ which is called the **quotient group** and written G/H. Hence

$$|G/H| = |G|/|H|.$$

We have shown above in Example 4.8 that the left and right cosets of A_3 in S_3 are identical and hence A_3 is a normal subgroup of S_3 (i.e. $A_3 \triangleleft S_3$) and the quotient group S_3/A_3 has $6/3 = 2$ elements. These are the cosets $id\,A_3 = \{id, \sigma, \sigma^2\}$ and $\tau A_3 = \{\tau, \sigma^2\tau, \sigma\tau\}$ and we have, for example,

$$(\tau A_3) * (\tau A_3) = \tau^2 A_3 = id\,A_3.$$

The presence of A_3 in this and other similar relations is redundant when trying to understand the structure of the quotient group as this is determined simply by the products of the elements in the original group. The two elements of S_3/A_3 are cosets, but S_3/A_3 is seen to be isomorphic to $\{id, \tau\}$. Thus

$$S_3/A_3 \cong C_2.$$

The formation of quotient groups is the group theoretic equivalent of division, though there is a lot of machinery which sits behind an equation such as $S_3/A_3 \cong C_2$. The orders of the groups behave just as in ordinary division, i.e. $6/3 = 2$, but the most important observation is that the group structure is preserved through the quotient process. Luckily, we don't usually have to go through detailed calculations like those above when dealing with groups of small order because the number of such groups (up to isomorphism) is rather limited.

Example 4.10 Any group of order 2 is isomorphic to C_2 and any group of order 3 is isomorphic to C_3. ♦

Example 4.11 Any subgroup of index 2 is normal. To see this we just need to note that there are two cosets associated with a subgroup $H \leq G$ of index 2, one of which will be $id\,H = H\,id$ with $|G|/2$ elements. Since the cosets divide G into two disjoint sets, the other coset cannot depend on whether we form the left or the right coset. Hence $H \lhd G$. ♦

If N is a normal subgroup of G, then by definition $gN = Ng$ for any $g \in G$. Equivalently

$$gNg^{-1} = N.$$

Thus if we take an element $\sigma \in N$ and form the product $g\sigma g^{-1}$ for any $g \in G$, the result will again be in the subgroup N. This gives an alternative way to test if a subgroup H is normal in G: compute $g\sigma g^{-1}$ for all $\sigma \in H$ and $g \in G$, $g \notin H$ and check that they all lie in H (there is no need to test elements g which are in H since H is a group and so $g\sigma g^{-1}$ must be in H if both g and σ are). We also have

Proposition 4.3 *For any subgroup H of G that contains N,*

$$N \lhd G \quad \textit{implies} \quad N \lhd H.$$

To see this we note that if $gNg^{-1} = N$ for all $g \in G$, then it is certainly true that this holds if G is replaced by any subgroup of G as this makes the condition less restrictive.

For two elements σ and τ in G, if there is an element $g \in G$ such that $g\sigma g^{-1} = \tau$, then we say that σ and τ are **conjugate** in G (not to be confused with the concept of a complex conjugate).

Example 4.12 Since $(x_1x_2)^{-1} = (x_2x_1)$ and

$$(x_1x_2)(x_1x_2x_3)(x_2x_1) = (x_1x_3x_2),$$

it follows that $(x_1x_2x_3)$ and $(x_1x_3x_2)$ are conjugate in S_3. ♦

The set of all elements of G that are conjugate to an element σ is known as the **conjugacy class** of σ and, much like cosets, they partition G into disjoint sets. If

N is a normal subgroup of G then we know that $g\sigma g^{-1}$ is in N for all $\sigma \in N$ and $g \in G$. Hence if $\sigma \in N$, then all conjugates of σ are in N. In other words any normal subgroup of G must be made up of a union of conjugacy classes of G. The condition $g\sigma g^{-1} = \tau$ is equivalent to $g\sigma = \tau g$ and so for an abelian group, τ must equal σ for them to be conjugate. Hence the conjugacy class of an element σ is just $\{\sigma\}$. This is consistent with the fact, mentioned above, that any subgroup of an abelian group is normal.

For the symmetric group S_n, it can be shown that the conjugacy classes are determined by cycle type.[4] In other words, conjugation in S_n doesn't affect cycle type and two elements of S_n are conjugate if and only if they have the same cycle type. Another way of saying the same thing is that conjugation in S_n is equivalent to a relabelling of the elements of the underlying set $\{x_1, \ldots, x_n\}$.

Example 4.13 The elements of S_3, together with their cycle types are:

$$\begin{array}{cccccc} id & (x_1x_2x_3) & (x_1x_3x_2) & (x_2x_3) & (x_1x_3) & (x_1x_2) \\ (1,1,1) & (3) & (3) & (2,1) & (2,1) & (2,1) \end{array}$$

and thus the three conjugacy classes of S_3 are

$$\{id\},\ \{(x_1x_2x_3), (x_1x_3x_2)\},\ \{(x_2x_3), (x_1x_3), (x_1x_2)\}.$$

This shows that $A_3 = \{id, (x_1x_2x_3), (x_1x_3x_2)\}$ is a normal subgroup of S_3, being the union of two of its conjugacy classes. However, $\{id, (x_2x_3)\}$ is not normal in S_3 since it does not contain the complete conjugacy class of (x_2x_3). ♦

We now turn to S_4, which has $4! = 24$ elements. Let's begin by listing them all, arranged by cycle type (i.e. conjugacy class):

$$\begin{array}{cl} (1,1,1,1) & id \\ (4) & (x_1x_2x_3x_4), (x_1x_2x_4x_3), (x_1x_3x_2x_4), \\ & (x_1x_3x_4x_2), (x_1x_4x_2x_3), (x_1x_4x_3x_2) \\ (3,1) & (x_1x_2x_3), (x_1x_3x_2), (x_1x_2x_4), (x_1x_4x_2), \\ & (x_1x_3x_4), (x_1x_4x_3), (x_2x_3x_4), (x_2x_4x_3) \\ (2,1,1) & (x_1x_2), (x_1x_3), (x_1x_4), (x_2x_3), (x_2x_4), (x_3x_4) \\ (2,2) & (x_1x_2)(x_3x_4), (x_1x_3)(x_2x_4), (x_1x_4)(x_2x_3). \end{array}$$

[4] https://kconrad.math.uconn.edu/blurbs/grouptheory/conjclass.pdf, Thm. 5.7.

Those elements with cycle type (4) have order 4, those with cycle type (3, 1) have order 3 and those with cycle type (2, 1, 1) or (2, 2) have order 2. From Lagrange's theorem, we know that the order of any proper subgroup H of S_4 must divide into 24 and so $|H|$ can be either 2, 3, 4, 6, 8 or 12. If we take any element σ from the above list that has order 2 or 3, then $\{id, \sigma\}$ is a subgroup, which will be isomorphic to C_2 or C_3, respectively. Such a subgroup cannot contain any of the conjugacy classes other than the identity (as the non-trivial conjugacy classes each have at least 3 elements) and so these subgroups are not normal in S_4. We can also show that there are four subgroups of order 6, each of which is isomorphic to S_3 and is obtained by fixing one of x_1, x_2, x_3 or x_4 and then considering S_3 acting on the remaining three elements. Again, none of these subgroups is normal in S_4.

The alternating group A_4, of order 12, consists of all the even permutations. Looking at the conjugacy classes above we see that it consists of the identity plus the 3-cycles (of which there are 8) plus the 3 permutations with cycle type (2, 2). As it is the union of conjugacy classes, or (see Example 4.11) simply because it has index 2 in S_4, we have $A_4 \lhd S_4$. We thus have $S_4/A_4 \cong C_2$.

Each of the 4-cycles generates a cyclic subgroup of order 4. If we take $\sigma = (x_1x_3x_2x_4)$, then $\sigma^2 = (x_1x_2)(x_3x_4)$ and $\sigma^3 = (x_1x_4x_2x_3)$ so, in particular, $(x_1x_3x_2x_4)$ and $(x_1x_4x_2x_3)$ generate the same subgroup. The three cyclic subgroups of order 4 are

$$\begin{aligned}
\langle\, (x_1x_3x_2x_4)\,\rangle &= \{id, (x_1x_3x_2x_4), (x_1x_2)(x_3x_4), (x_1x_4x_2x_3)\},\\
\langle\, (x_1x_2x_3x_4)\,\rangle &= \{id, (x_1x_2x_3x_4), (x_1x_3)(x_2x_4), (x_1x_4x_3x_2)\},\\
\langle\, (x_1x_2x_4x_3)\,\rangle &= \{id, (x_1x_2x_4x_3), (x_1x_4)(x_2x_3), (x_1x_3x_4x_2)\}.
\end{aligned} \tag{4.3}$$

None of these are the union of conjugacy classes and so they are not normal subgroups of S_4. Since a 4-cycle is an odd permutation, we infer that even though 4 divides into 12, these cyclic subgroups of S_4 are not subgroups of A_4.

As we saw earlier, cyclic groups are isomorphic to groups of rotations of regular polygons. If we add a reflection into the group we get the full symmetry group of a regular polygon; these are called the dihedral groups. The full symmetry group of a square is illustrated in Fig. 4.2 in which the top four images show the effect of the identity and then rotations anticlockwise through 90°, 180° and 270°. The lower four images are the reflections of the top four about the horizontal symmetry axis of the square. The letter G serves to differentiate the 8 possible outcomes of these operations. In the geometric context, dihedral groups are usually labelled according to the regular polygon with which they are associated so that the group illustrated in Fig. 4.2 would be referred to as D_4, but in the algebraic context the dihedral groups are usually labelled according to their order. We follow this latter convention so that the full symmetric group of a square is labelled D_8.

The figure shows that D_8 is generated by a rotation through 90° and a reflection. If we label the vertices of the square with x_1, x_2, x_3 and x_4, then a rotation through 90° is represented in S_4 by a 4-cycle, which generates one of the cyclic groups in (4.3), different cyclic groups corresponding to different ways of labelling the

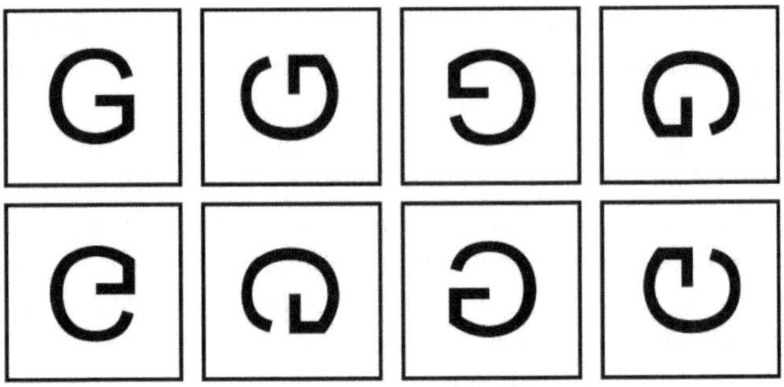

Fig. 4.2 The dihedral group D_8 acting on a square. From left to right, the top four images show the effect of the identity and then rotations anticlockwise through 90°, 180° and 270°. The lower four images are the reflections of the top four about the horizontal symmetry axis of the square. If we label the vertices of the square x_1, x_2, x_3, x_4 in anticlockwise fashion, then the upper images correspond, in order, to the permutations id, $(x_1x_2x_3x_4)$, $(x_1x_3)(x_2x_4)$, $(x_1x_4x_3x_2)$ and the lower images to $(x_1x_4)(x_2x_3)$, (x_1x_3), $(x_1x_2)(x_3x_4)$, (x_2x_4). With $\sigma = (x_1x_2x_3x_4)$ and $\tau = (x_1x_3)$, these correspond, respectively, to $id, \sigma, \sigma^2, \sigma^3$ along the top and $\tau\sigma, \tau, \tau\sigma^3, \tau\sigma^2$ along the bottom

vertices. Similarly, a reflection is represented in S_4 by a transposition, or a product of transpositions. If we imagine the vertices of the square labeled x_1, x_2, x_3, x_4, in an anticlockwise fashion (corresponding to the second cyclic group in (4.3)), then (x_1x_3), (x_2x_4) represent refections about the diagonals and $(x_1x_2)(x_3x_4)$, $(x_1x_4)(x_2x_3)$ represent reflections about the vertical and horizontal symmetry axes, respectively.

We thus find that there are three subgroups of S_4 that are isomorphic to D_8, all of the form

$$\langle\sigma, \tau\rangle = \{id, \sigma, \sigma^2, \sigma^3, \tau, \tau\sigma, \tau\sigma^2, \tau\sigma^3\}, \quad \sigma^4 = \tau^2 = id, \ \sigma\tau = \tau\sigma^3$$

in which the element σ can be thought of as representing a rotation, while τ represents a reflection. We will label the group generated by $\sigma = (x_1x_3x_2x_4)$ and $\tau = (x_1x_2)$ as $D_8^{(1)}$, that generated by $\sigma = (x_1x_2x_3x_4)$ and $\tau = (x_1x_3)$ as $D_8^{(2)}$ and that generated by $\sigma = (x_1x_2x_4x_3)$ and $\tau = (x_1x_4)$ as $D_8^{(3)}$. Explicitly, in the case of $D_8^{(1)}$, $\{id, \sigma, \sigma^2, \sigma^3, \tau, \tau\sigma, \tau\sigma^2, \tau\sigma^3\}$ is, in the same order,

$$\begin{aligned} D_8^{(1)} = \{&id, (x_1x_3x_2x_4), (x_1x_2)(x_3x_4), (x_1x_4x_2x_3), \\ &(x_1x_2), (x_1x_4)(x_2x_3), (x_3x_4), (x_1x_3)(x_2x_4)\}. \end{aligned} \tag{4.4}$$

None of $D_8^{(1)}$, $D_8^{(2)}$ or $D_8^{(3)}$ is normal in S_4 as they are not a union of conjugacy classes and they are not subgroups of A_4 since they contain odd permutations (this also follows from the fact that 8 does not divide into 12).

There are four other subgroups of S_4, all of which are isomorphic to the **Klein 4-group**, V_4, which can be thought of as the full symmetry group of a non-square rectangle and is the smallest non-cyclic group. We have

$$V_4 = \{id, (x_1x_2)(x_3x_4), (x_1x_3)(x_2x_4), (x_1x_4)(x_2x_3)\},$$

which consists of even permutations only and hence $V_4 \subset A_4$. Since V_4 is a union of two conjugacy classes we see that $V_4 \lhd S_4$ and looking at the orders of the groups shows us that $|S_4/V_4| = 6$. There are only two groups of order 6 (up to isomorphism), the cyclic group C_6, which is abelian, and S_3 which is non-abelian. It is not difficult to show that

$$S_4/V_4 \cong S_3.$$

Since V_4 is normal in S_4, it is normal in any subgroup of S_4 that contains it; see Proposition 4.3. Hence $V_4 \lhd A_4$ and

$$A_4/V_4 \cong C_3.$$

The other three subgroups which are isomorphic to V_4, are

$$\begin{aligned}
&\{id, (x_1x_2), (x_3x_4), (x_1x_2)(x_3x_4)\},\\
&\{id, (x_1x_3), (x_2x_4), (x_1x_3)(x_2x_4)\},\\
&\{id, (x_1x_4), (x_2x_3), (x_1x_4)(x_2x_3)\},
\end{aligned} \tag{4.5}$$

none of which is a subset of A_4 and none of which is normal in S_4. These three groups and V_4 are all of the form

$$\{id, \sigma, \tau, \sigma\tau\}, \quad \sigma^2 = \tau^2 = id, \; \sigma\tau = \tau\sigma.$$

The groups S_4 and A_4 are examples of groups that have proper subgroups which are normal. In group theory, a particularly important class of finite groups are the **simple** groups, which are defined as those which have no proper normal subgroups. In other words, a group is said to be simple if its only normal subgroups are $\{id\}$ and itself. Cyclic groups of prime order are all simple, for example. The reason for their importance is that if a group has a proper normal subgroup, then a smaller non-trivial group can be obtained by forming the corresponding quotient group. In this way, all finite groups can be broken down into a set of simple groups and thus simple groups can be thought of as the building blocks out of which all other finite groups are constructed. The classification of all finite simple groups was a major mathematical endeavour, which spanned about 50 years, culminating in the final pieces of the jigsaw appearing in print in the early years of the twenty-first century. As we shall see, the fact that A_5 is simple (which we will establish in Chap. 7) plays a crucial role in the theory of the quintic.

Field Extensions

5

In Sect. 3.1 we introduced the concept of a field, a set of numbers which allows us to perform the operations of addition, subtraction, multiplication and division (except by zero) without leaving the set. When finding the roots of polynomials we need to extract roots: square roots, cube roots and so on, and these operations are not included in the definition of a field. If we start with the set of rationals $\mathbb{Q}$ then as soon as we allow square roots we have a problem, since numbers like $\sqrt{2}$ are not rational. So we need to enlarge the field we are working with if we want to include the roots of equations such as x^2-2. We can of course just take the complex numbers $\mathbb{C}$, since the fundamental theorem of algebra tells us that all roots of all polynomials are in $\mathbb{C}$, but that doesn't tell us anything new. In this chapter we try and shed light on the question, what is the smallest field that contains all the roots of a given polynomial? For an nth degree polynomial there are at most n distinct roots, so is it possible to find a field that helps to characterise them in some way?

The basic idea is to start with the base field F and extend it by adding new numbers. By adding a number, we mean enlarging the field to include any number that can be generated using the standard arithmetic operations on elements of F and the new number. More precisely, if F is a subfield of $\mathbb{C}$ and $S = \{\alpha_1, \ldots, \alpha_m\}$ is a finite subset of $\mathbb{C}$, we write $F(S)$ or $F(\alpha_1, \ldots, \alpha_m)$ for the **field extension** which is the subfield of $\mathbb{C}$ that is generated by (finitely many of) the operations $+$, $-$, $\times$, $\div$ applied to all the elements of F and S. Thus $\mathbb{Q}(\sqrt{2})$ is a field which contains, as well as all the rationals, numbers such as $1+\sqrt{2}$, $1/(1-\sqrt{2})$, $(\sqrt{2})^5$, and so on. We will show how to characterise this field in Example 5.3 below. Equivalently, $F(S)$ is the smallest subfield of $\mathbb{C}$ which contains both F and S. Clearly, if all the elements of S are already in F, then $F(S) = F$. If $K = F(S)$ then we also use the notation K/F ("K over F") to denote the field extension.

If the set S contains more than one member, we can adjoin the elements in turn: in other words $F(\alpha_1, \ldots, \alpha_m) = F(\alpha_1)(\alpha_2)\ldots(\alpha_m)$ and we have

$$F \subseteq F(\alpha_1) \subseteq F(\alpha_1, \alpha_2) \subseteq \ldots \subseteq F(\alpha_1, \ldots, \alpha_m).$$

C. M. Linton, *Quintic Equations and How to Solve Them*,
https://doi.org/10.1007/978-3-032-01658-4_5

The elements can be adjoined in any order, though the intermediate fields in the chain may then differ. Thus $\mathbb{Q}(\sqrt{2}, \sqrt{3}) = \mathbb{Q}(\sqrt{2})(\sqrt{3})$ and

$$\mathbb{Q} \subseteq \mathbb{Q}(\sqrt{2}) \subseteq \mathbb{Q}(\sqrt{2}, \sqrt{3})$$

but equally, $\mathbb{Q}(\sqrt{2}, \sqrt{3}) = \mathbb{Q}(\sqrt{3})(\sqrt{2})$ and

$$\mathbb{Q} \subseteq \mathbb{Q}(\sqrt{3}) \subseteq \mathbb{Q}(\sqrt{2}, \sqrt{3}).$$

An important property of a field extension is its degree, which can be thought of as the number of independent parameters that are required to characterise its elements. For example, in a way that we will make precise shortly, the complex numbers, considered as an extension of the real numbers, have degree 2 because any complex number is defined in terms of two real numbers. To give a formal definition of degree we first define what it means for a set to be a basis for a field extension.

Given an extension K/F, we can try and find a set $\{a_1, \dots, a_n\}$ of elements of K which has two properties:

- $\{a_1, \dots, a_n\}$ **spans** K, by which we mean that any element of K can be written in the form of a linear combination $\mu_1 a_1 + \dots + \mu_n a_n$, where each of the coefficients $\mu_i \in F$,
- $\{a_1, \dots, a_n\}$ is **linearly independent**, by which we mean that $\mu_1 a_1 + \dots + \mu_n a_n = 0$ implies that $\mu_1 = \dots = \mu_n = 0$.

If such a set exists then we say that $\{a_1, \dots, a_n\}$ is a **basis** for K/F and the **degree of the field extension** is $[K : F] = n$. It is a standard theorem of linear algebra that any set that is linearly independent and spans K will have the same number of elements, so the degree is uniquely determined.

Proposition 5.1 *If K/L is an extension of degree n and L/F is an extension of degree m, then K is an extension of F and*

$$[K : F] = [K : L][L : F] = nm.$$

This follows because if the set $\{a_1, \dots, a_n\}$ is a basis for K over L and $\{b_1, \dots, b_m\}$ is a basis for L over F, then $\{a_i b_j\}$, $i = 1, \dots, n$, $j = 1, \dots, m$ is a basis for K over F.

Example 5.1 Any complex number is of the form $a + ib$ where a and b are real. Hence $\mathbb{C} = \mathbb{R}(\mathrm{i})$, $\{1, \mathrm{i}\}$ is a basis for $\mathbb{C}/\mathbb{R}$, and $[\mathbb{C} : \mathbb{R}] = 2$. ♦

Example 5.2 $[K : K] = 1$ since $\{1\}$ serves as a basis. The converse is also true: if $[K : F] = 1$, then $K = F$. ♦

Example 5.3 Consider the extension $\mathbb{Q}(\sqrt{2})$. This is the field that is generated by the operations $+, -, \times, \div$ applied to all rational numbers and $\sqrt{2}$. Equivalently, we can think of it as the smallest field which contains both $\mathbb{Q}$ and $\sqrt{2}$. Consider the set

$$K = \{a + b\sqrt{2}, a, b \in \mathbb{Q}\},$$

which contains $\mathbb{Q}$ (set $b = 0$) and $\sqrt{2}$ (set $a = 0$ and $b = 1$). It is clearly closed under the operations $+, -, \times$ and this is true also for division since

$$\frac{1}{a + b\sqrt{2}} = \frac{a - b\sqrt{2}}{a^2 - 2b^2} = \frac{a}{a^2 - 2b^2} - \frac{b}{a^2 - 2b^2}\sqrt{2},$$

noting that the irrationality of $\sqrt{2}$ ensures that the denominators do not vanish unless $a = b = 0$. Hence K is a field which contains $\mathbb{Q}$ and $\sqrt{2}$; in other words $\mathbb{Q}(\sqrt{2}) \subseteq K$. But any field that contains $\mathbb{Q}$ and $\sqrt{2}$ must contain all numbers of the form $a + b\sqrt{2}$ and hence $K \subseteq \mathbb{Q}(\sqrt{2})$. The only possibility is that $\mathbb{Q}(\sqrt{2}) = K$.

A basis for $\mathbb{Q}(\sqrt{2})$ is $\{1, \sqrt{2}\}$ and $[\mathbb{Q}(\sqrt{2}) : \mathbb{Q}] = 2$. The same argument shows that $[\mathbb{Q}(\sqrt{n}) : \mathbb{Q}] = 2$ whenever n is a non-square positive integer (in which case $\sqrt{n}$ is irrational). ♦

Example 5.4 Consider the extension $\mathbb{Q}(\sqrt{2}, \sqrt{3})$, an example that we will return to often. We claim that this is equivalent to $\mathbb{Q}(\gamma)$, where $\gamma = \sqrt{2}+\sqrt{3}$. Clearly $\gamma \in \mathbb{Q}(\sqrt{2}, \sqrt{3})$ and therefore $\mathbb{Q}(\gamma) \subseteq \mathbb{Q}(\sqrt{2}, \sqrt{3})$. Now note that

$$\frac{1}{\sqrt{2}+\sqrt{3}} = \frac{\sqrt{2}-\sqrt{3}}{(\sqrt{2}+\sqrt{3})(\sqrt{2}-\sqrt{3})} = \sqrt{3}-\sqrt{2}$$

and so $\sqrt{3}-\sqrt{2} \in \mathbb{Q}(\gamma)$. Since both $\sqrt{3} \pm \sqrt{2} \in \mathbb{Q}(\gamma)$, it follows that their sum and difference, and hence $\sqrt{2}$ and $\sqrt{3}$, are both in $\mathbb{Q}(\gamma)$ and so $\mathbb{Q}(\sqrt{2}, \sqrt{3}) \subseteq \mathbb{Q}(\gamma)$. Hence $\mathbb{Q}(\sqrt{2}+\sqrt{3}) = \mathbb{Q}(\sqrt{2}, \sqrt{3})$.

We can determine a basis for $\mathbb{Q}(\sqrt{2}, \sqrt{3})$ using Proposition 5.1. Let $F = \mathbb{Q}$, $L = \mathbb{Q}(\sqrt{2})$ and $K = \mathbb{Q}(\sqrt{2}, \sqrt{3})$. Then a basis for L/F is $\{1, \sqrt{2}\}$ and a basis for K/L is $\{1, \sqrt{3}\}$. Hence a basis for $\mathbb{Q}(\sqrt{2}, \sqrt{3})$ is

$$\{1, \sqrt{2}, \sqrt{3}, \sqrt{6}\}$$

and the degree of $\mathbb{Q}(\sqrt{2}, \sqrt{3})/\mathbb{Q}$ is 4. ♦

Fortunately, there is an easier way to determine the degree of a field extension than having to work out a basis, and it involves minimal polynomials. Consider the extension $F(\alpha)/F$, where α is algebraic over F. Let p be the minimal polynomial of α and assume it has degree n. We claim that the set $\{1, \alpha, \ldots, \alpha^{n-1}\}$ is then a basis for $F(\alpha)/F$. To establish this we need to check that the set $\{1, \alpha, \ldots, \alpha^{n-1}\}$ spans $F(\alpha)$ and that it is linearly independent. Remember that $F(\alpha)$ is the set of numbers that can be obtained from the operations $+, -, \times, \div$ applied to all the elements of F, together with α. All such numbers will be of the form $f(\alpha)/g(\alpha)$, where $f, g \in F[x]$ and $g(\alpha) \neq 0$.

Any polynomial in α can be written as a linear combination of $\{1, \alpha, \ldots, \alpha^{n-1}\}$ since any powers of α higher than $n-1$ can be replaced by a linear combination of powers less than n by using the fact that $p(\alpha) = 0$. Since $g(\alpha) \neq 0$, we know that p is not a factor of g. Moreover, since p is a minimal polynomial, it is irreducible (Proposition 3.2) and hence has no factors in $F[x]$ other than 1 and itself. It follows that the highest common factor of g and p is 1. We now appeal to a result known as Bézout's identity which implies that under these circumstances we can find polynomials $h, k \in F[x]$ such that $hg + kp = 1$. Evaluating this at α shows that $h(\alpha)g(\alpha) = 1$ and hence that division by $g(\alpha)$ can be replaced by multiplication by a different polynomial in $F[x]$, namely h. This is sufficient to establish that $\{1, \alpha, \ldots, \alpha^{n-1}\}$ spans $F(\alpha)$. To prove linear independence, set

$$\mu_{n-1}\alpha^{n-1} + \cdots + \mu_1\alpha + \mu_0 = 0.$$

All the coefficients $\mu_0, \ldots, \mu_{n-1}$ must be equal to zero otherwise we would have a polynomial in $F[x]$ that is zero at α with a degree less than n, contradicting the fact that the minimal polynomial of α has degree n.

We have established the crucial result that:

Proposition 5.2 *If α is algebraic over F, then $[F(\alpha) : F]$, the degree of the extension $F(\alpha)/F$, is equal to the degree of the minimal polynomial of α over F.*

Example 5.5 Consider the extension $\mathbb{Q}(\zeta_p)$, where p is prime and ζ_p is a pth root of unity other than 1. From Example 3.12 we know that the minimal polynomial of ζ_p over $\mathbb{Q}$ is $x^{p-1} + \cdots + x + 1$, which has degree $p - 1$. Hence

$$[\mathbb{Q}(\zeta_p) : \mathbb{Q}] = p - 1. \qquad \blacklozenge$$

Example 5.6 The field extension $\mathbb{Q}(\sqrt{2}, \sqrt{3})$ can be thought of as $\mathbb{Q}(\sqrt{2})(\sqrt{3})$ and since the minimal polynomial of $\sqrt{2}$ over $\mathbb{Q}$ is $x^2 - 2$ (of degree 2) and likewise for $\sqrt{3}$ (over $\mathbb{Q}(\sqrt{2})$), we see from Proposition 5.1 that the degree of the extension $\mathbb{Q}(\sqrt{2}, \sqrt{3})$ is 4 as shown in Example 5.4.

Since $\mathbb{Q}(\sqrt{2}, \sqrt{3}) = \mathbb{Q}(\sqrt{2} + \sqrt{3})$, it follows that the minimal polynomial of $\sqrt{2} + \sqrt{3}$ over $\mathbb{Q}$ has degree 4. We can calculate

$$(\sqrt{2} + \sqrt{3})^2 = 5 + 2\sqrt{6}, \qquad (\sqrt{2} + \sqrt{3})^4 = 49 + 20\sqrt{6}$$

and so $\sqrt{2} + \sqrt{3}$ is a root of $x^4 - 10x^2 + 1$, which is therefore its minimal polynomial.

However, the minimal polynomial of $\sqrt{2} + \sqrt{3}$ over $\mathbb{Q}(\sqrt{2})$, or $\mathbb{Q}(\sqrt{3})$, has degree 2. Since

$$(\sqrt{2} + \sqrt{3})^2 = 5 + 2\sqrt{2}\sqrt{3}$$

we can compute the minimal polynomial over $\mathbb{Q}(\sqrt{2})$ as $x^2 - 2\sqrt{2}x - 1$ and over $\mathbb{Q}(\sqrt{3})$ as $x^2 - 2\sqrt{3}x + 1$. $\blacklozenge$

Example 5.7 Since $\{1, \sqrt{2}, \sqrt{3}, \sqrt{6}\}$ is a basis for $\mathbb{Q}(\sqrt{2}, \sqrt{3})$, we see that $\{1, \sqrt{2}\}$, $\{1, \sqrt{3}\}$, $\{1, \sqrt{6}\}$, each form a basis for a subfield of $\mathbb{Q}(\sqrt{2}, \sqrt{3})$. We can illustrate all the subfields of $\mathbb{Q}(\sqrt{2}, \sqrt{3})$ in a diagram in which, reading downwards, the solid lines connect fields with their subfields and the numbers indicate the degree of the extension in each case.

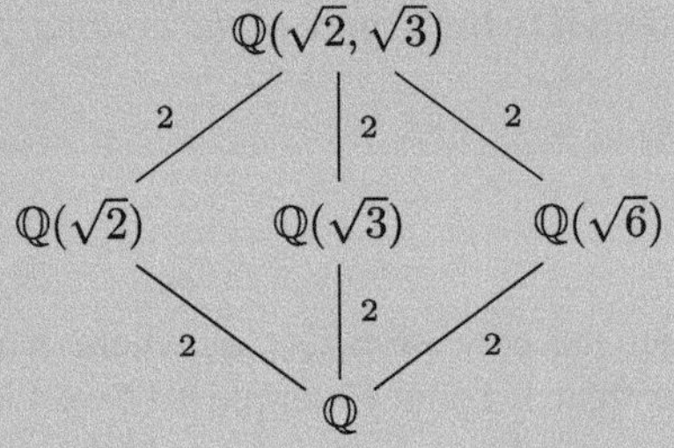

How do we know that there are no other intermediate fields between $\mathbb{Q}$ and $\mathbb{Q}(\sqrt{2}, \sqrt{3})$? At this stage, we don't. It will, however, be a consequence of the fundamental theorem of Galois theory, which we will get to in the next chapter. ♦

Example 5.8 If we have a sequence of distinct primes $p_1, \ldots, p_n$, then $\sqrt{p_n}$ is not an element of $\mathbb{Q}(\sqrt{p_1}, \ldots, \sqrt{p_{n-1}})$. It follows that the minimal polynomial of $\sqrt{p_n}$ over $\mathbb{Q}(\sqrt{p_1}, \ldots, \sqrt{p_{n-1}})$ is $x^2 - p_n = 0$ and hence that

$$[\mathbb{Q}(\sqrt{p_1}, \ldots, \sqrt{p_n}) : \mathbb{Q}(\sqrt{p_1}, \ldots, \sqrt{p_{n-1}})] = 2$$

from which

$$[\mathbb{Q}(\sqrt{p_1}, \ldots, \sqrt{p_n}) : \mathbb{Q}] = 2^n.$$

The fact that $[\mathbb{Q}(\sqrt{2}, \sqrt{3}) : \mathbb{Q}] = 4$, as shown in Example 5.4, is a special case. ♦

Example 5.9 Consider $\mathbb{Q}(\sqrt[3]{2}, \omega)$, where ω is the cube root of unity defined in (3.1). Since $x^3 - 2 = 0$ is the minimal polynomial of $\sqrt[3]{2}$ over $\mathbb{Q}$, we have

$$[\mathbb{Q}(\sqrt[3]{2}) : \mathbb{Q}] = 3.$$

The minimal polynomial of ω over $\mathbb{Q}$ is $x^2 + x + 1$ and since $\mathbb{Q}(\sqrt[3]{2}) \subset \mathbb{R}$ and $x^2 + x + 1$ has no real roots, it is also the minimal polynomial of ω over $\mathbb{Q}(\sqrt[3]{2})$. Hence

$$[\mathbb{Q}(\sqrt[3]{2}, \omega) : \mathbb{Q}(\sqrt[3]{2})] = 2.$$

Combining these shows that

$$[\mathbb{Q}(\sqrt[3]{2}, \omega) : \mathbb{Q}] = 6.$$

The other roots of $x^3 - 2$ are $\omega\sqrt[3]{2}$ and $\omega^2\sqrt[3]{2}$ and so we also have

$$[\mathbb{Q}(\omega\sqrt[3]{2}) : \mathbb{Q}] = [\mathbb{Q}(\omega^2\sqrt[3]{2}) : \mathbb{Q}] = 3.$$

The various subfields of $\mathbb{Q}(\sqrt[3]{2}, \omega)$ are illustrated via the following diagram:

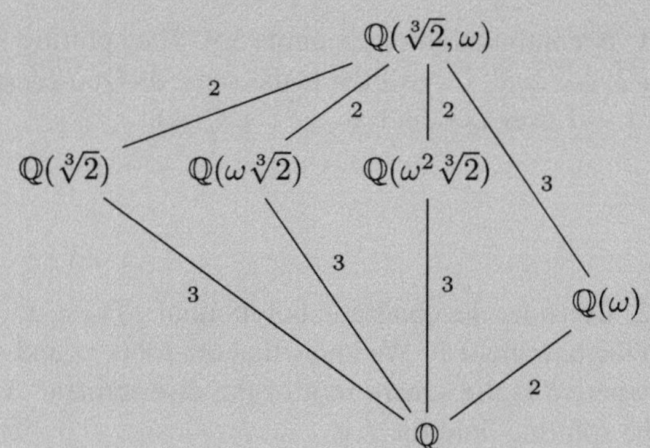

♦

We have described above how fields can be extended by adjoining new numbers and we now turn to the case where those new numbers are in fact the roots of a given polynomial $f \in F[x]$. We would like to construct a field K which is an extension of the base field F and which contains all the roots of f. We say that

$f = x^n + a_{n-1}x^{n-1} + \cdots + a_1 x + a_0$ **splits** in the extension K/F if it can be written as a product of linear factors, i.e. if we can write

$$f = (x - x_1) \ldots (x - x_n), \qquad x_1, \ldots, x_n \in K.$$

Thus f splits in K precisely when K contains all the roots of f. Clearly f splits in $F(x_1, \ldots, x_n)$ and this is the smallest field that contains both F and the roots of f. Hence, f splits in $K = F(x_1, \ldots, x_n)$ but not in any proper subfield of K and we call K the **splitting field** for f over F.

If all the roots of f are in F, then the splitting field is simply F itself. If some of the roots of f are in F, then we don't need to worry about those roots that are in F when constructing the splitting field. Even when none of the roots of f are in F, it may well be the case that we only need to adjoin some subset of the roots in order to obtain the splitting field, as in the next set of examples.

Example 5.10 The splitting field for $x^2 - 2$ over $\mathbb{Q}$ is $\mathbb{Q}(\sqrt{2}, -\sqrt{2})$ which is simply $\mathbb{Q}(\sqrt{2})$. The splitting field for the same polynomial over $\mathbb{R}$ is $\mathbb{R}$. ♦

Example 5.11 A continuation of Example 5.9. The splitting field of $x^3 - 2$ over $\mathbb{Q}$ is $\mathbb{Q}(\sqrt[3]{2}, \omega\sqrt[3]{2}, \omega^2\sqrt[3]{2})$ which is the same as $\mathbb{Q}(\sqrt[3]{2}, \omega)$. The splitting field for $x^2 + x + 1$ over $\mathbb{Q}$ is $\mathbb{Q}(1, \omega, \omega^2) = \mathbb{Q}(\omega)$. ♦

Example 5.12 Consider the quadratic polynomial $f(x) = x^2 + bx + c$ with coefficients in the base field F. We know that the roots x_1 and x_2 are given by $(-b \pm \delta)/2$, where δ is the square root of the discriminant $\Delta = b^2 - 4c$. It follows that the splitting field for f is

$$F(x_1, x_2) = F(\delta).$$

If $\delta \in F$, then $F(\delta) = F$ and f is reducible; it splits into linear factors in F, namely

$$f = \tfrac{1}{4}(2x + b + \delta)(2x + b - \delta).$$

(continued)

Example 5.12 (continued)

If $\delta \notin F$, then the minimal polynomial for δ over F is $x^2 - \Delta$ (remember that Δ is always in the base field) and so $[F(\delta) : F] = 2$ and $\{1, \delta\}$ is a basis for $F(\delta)/F$. ♦

We now make some important observations about the degree of the extension K/F, where K is the splitting field of F. If x_1 is a root of an nth degree equation in $F[x]$, then the minimal polynomial of x_1 over F has degree at most n. Hence $[F(x_1) : F] \leq n$. The polynomial f has at least one linear factor over the field $F(x_1)$, namely $(x - x_1)$, and so the minimal polynomial of x_2 over $F(x_1)$ has degree at most $n - 1$. Hence $[F(x_1, x_2) : F(x_1)] \leq n - 1$. Repeating this process shows that:

Proposition 5.3 *If K is the splitting field of $f \in F[x]$, then*

$$[K : F] \leq n!$$

In the case of an irreducible polynomial $f \in F[x]$, we know that f is the minimal polynomial over F of each of its roots and so $[F(x_1) : F] = n$ in that case. Thus:

Proposition 5.4 *If f is irreducible over F, with splitting field K, then $[K : F]$ is a multiple of n.*

On the other hand, if f is reducible over F, we know that the degree of the minimal polynomial of x_1 over F is strictly less than n and so $[K : F] < n!$. Thus:

Proposition 5.5 *If K is the splitting field of $f \in F[x]$ and $[K : F] = n!$, then f is irreducible over F.*

The converse is false, as the following example illustrates. Irreducibility does not imply that the degree of the splitting field is equal to $n!$, where n is the degree of the polynomial.

Example 5.13 The splitting field for $x^4 - 2$ over $\mathbb{Q}$ is $\mathbb{Q}(\alpha, -\alpha, i\alpha, -i\alpha)$, with $\alpha = \sqrt[4]{2}$, which is the same as $\mathbb{Q}(\alpha, i)$. Since the minimal polynomial of α over $\mathbb{Q}$ is $x^4 - 2$ we have $[\mathbb{Q}(\alpha) : \mathbb{Q}] = 4$. Also, $[\mathbb{Q}(\alpha, i) : \mathbb{Q}(\alpha)] = 2$ because the minimal polynomial of i over $\mathbb{Q}(\alpha)$ is $x^2 + 1$. Hence

$$[\mathbb{Q}(\alpha, i) : \mathbb{Q}] = 8.$$

A basis for $[\mathbb{Q}(\alpha, i) : \mathbb{Q}]$ is $\{1, \alpha, \alpha^2, \alpha^3, i, i\alpha, i\alpha^2, i\alpha^3\}$.

In this example, the degree of the splitting field over $\mathbb{Q}$ is 8, which is a multiple of 4 as it must be since $x^4 - 2$ is irreducible over $\mathbb{Q}$. However, even though $x^2 - 4$ is irreducible over $\mathbb{Q}$, the degree of its splitting field is not equal to 4!. ♦

Example 5.14 Let $f = x^p - 2 \in \mathbb{Q}[x]$, where p is prime. The solutions are $\zeta_p^m \sqrt[p]{2}$, $m = 0, 1 \ldots p - 1$, where ζ_p is a non-trivial pth root of unity. The splitting field is thus $K = \mathbb{Q}(\zeta_p, \sqrt[p]{2})$. The minimal polynomials over $\mathbb{Q}$ of ζ_p and $\sqrt[p]{2}$ have degrees $p - 1$ and p, respectively. Hence

$$[K : \mathbb{Q}] \leq p(p - 1).$$

But we also know that $\mathbb{Q}(\zeta_p)$ and $\mathbb{Q}(\sqrt[p]{2})$ are subfields of K and hence we know that $[K : \mathbb{Q}]$ is divisible by both p and $p - 1$. Since these numbers are co-prime, we must have

$$[\mathbb{Q}(\zeta_p, \sqrt[p]{2}) : \mathbb{Q}] = p(p - 1).$$

As a by-product, this shows that $[\mathbb{Q}(\zeta_p, \sqrt[p]{2}) : \mathbb{Q}(\sqrt[p]{2})] = p - 1$. (Examples 5.9 and 5.11 treat the case $p = 3$). ♦

Example 5.15 Let $\alpha = \sqrt{2 + \sqrt{2}}$. Since $\alpha^2 = 2 + \sqrt{2}$ and $\alpha^4 = 6 + 4\sqrt{2}$, we see that the minimal polynomial of α over $\mathbb{Q}$ is $x^4 - 4x^2 + 2$ and hence $[\mathbb{Q}(\alpha) : \mathbb{Q}] = 4$. The roots of the biquadratic $x^4 - 4x^2 + 2$ are easily calculated to be the four numbers $\sqrt{2 \pm \sqrt{2}}$, $-\sqrt{2 \pm \sqrt{2}}$ and since

$$\frac{1}{\sqrt{2 + \sqrt{2}}} = \frac{\sqrt{2 - \sqrt{2}}}{\sqrt{2}}, s$$

(continued)

Example 5.15 (continued)

we have $\sqrt{2-\sqrt{2}} = (\alpha^2 - 2)/\alpha$. Thus all of the roots are contained in $\mathbb{Q}(\alpha)$, which is therefore the splitting field for $x^4 - 4x^2 + 2$ over $\mathbb{Q}$. ♦

Example 5.16 Let $f(x) = x^4 + 4$. From Example 3.3, f is reducible over $\mathbb{Q}$:

$$x^4 + 4 = (x^2 + 2x + 2)(x^2 - 2x + 2)$$

and we can compute the four roots as $1 \pm \mathrm{i}, \ -1 \pm \mathrm{i}$. Hence the splitting field of f is simply $\mathbb{Q}(\mathrm{i})$ with degree 2. This is not a multiple of 4 as it need not be, since f is reducible. ♦

In Examples 5.10, 5.15 and 5.16 above, the splitting field is obtained simply by adjoining one of the roots of the polynomial to the base field $\mathbb{Q}$. But in Examples 5.13 and 5.14, adjoining a single root is insufficient to get the splitting field. This is suggestive of the fact that the structure of the splitting field contains important information about the roots of the original polynomial.

Galois Theory

6

We have mentioned Galois' name in the context of the development of group theory, but his contributions to the theory of equations were so profound that the theory that now underpins this branch of mathematics bears his name. He also led a rather interesting, albeit, tragically brief life and so it seems appropriate to say a few words about him here (based on [60] and [52]).

Évariste Galois was a revolutionary, both in terms of his mathematics and also through his association with some of the more extreme elements within the French republican movement during the politically turbulent times in which he lived. Against advice he took the entrance exam for the École Polytechnique a year early—he failed. In 1829, aged 17, he took the exam again (just a few days after the suicide of his father). Again, he failed, and had to settle for a place at the less prestigious École Normale, by all accounts extremely bitter about his treatment. He was expelled in December 1830 for writing a letter publicly criticising the school's Director and immediately joined the Artillery of the National Guard, a republican division that was abolished by royal decree that same month. Not long afterwards, following a republican banquet in May 1831, he was arrested for threatening the king's life but acquitted, largely on the grounds of his youth. He was arrested again, on Bastille Day, for wearing the uniform of the disbanded Artillery Guard (and armed to the teeth) and he was sentenced to 6 months incarceration.

In mathematics, Galois showed flair as a teenager but was not one to bother mastering all the basics. He published a short article on continued fractions when he was just 17 and soon afterwards submitted a memoir on the solvability of equations of prime degree to the French Academy, at the time one of the world's most preeminent scientific institutions. Cauchy was appointed to review it. For reasons that are not entirely clear, Cauchy did not present Galois' work to the Academy and instead appears to have encouraged Galois to submit a revised version for the Grand Prize in Mathematics, which he duly did prior to the deadline of 1 March 1830. This time his work was in the hands of Joseph Fourier. Misfortune, whether self-inflicted

C. M. Linton, *Quintic Equations and How to Solve Them*,
https://doi.org/10.1007/978-3-032-01658-4_6

or otherwise, was never far away from Galois and Fourier's death in May of that year resulted in the loss of the manuscript.

Galois did still manage to get some of his ideas into print; three brief articles were published in 1830 but no-one took much notice of the work of an 18-year-old student. In October 1831, while in prison, Galois was notified of the rejection, this time by Siméon Denis Poisson, of his latest submission to the Academy. Poisson wrote

> We have made every effort to understand M. Galois's proofs. His argument is neither sufficiently clear nor sufficiently developed to allow us to judge its rigour; it is not even possible for us to give an idea of this paper. The author claims that the propositions contained in his manuscript are a part of a general theory which has rich application. Often different parts of a theory clarify each other and can be more easily understood when taken together than when taken in isolation. One should rather wait to form a more definite opinion, therefore, until the author publishes a more complete account of his work.

Galois' anger at the failure of the Academy to appreciate his contributions boiled over and he decided that he would instead get his ideas into print through the services of a friend, Auguste Chevalier. In fact, one of the first things he wrote was a diatribe against his treatment by the academic elite.

Galois was released from prison in early 1832, but this was not to be the start of a new, more settled, period in his life. On the morning of 30 May, 1832, aged 21, Galois fought a duel the causes of which seem most likely to have been related to a broken love affair. He was shot in the stomach and subsequently died. The night before, he wrote a letter to his friend Chevalier expanding on his ideas and he annotated and made corrections to some of his papers, included the one Poisson had rejected. A translation of this letter into English can be found in [20, Chapter 15.D] and a detailed analysis of its contents is given in [52, Chapter III]. Galois concluded his letter with the hope that "there will be people who will find it to their advantage to decipher all this mess". This hope was certainly realised, though it took some time. His papers eventually ended up with Joseph Liouville, who published them in 1846. It is remarkable that someone who spent only a few years studying mathematics and whose life could hardly be described as straightforward was capable of such original ideas with such far reaching consequences.

It is tempting, also, to be critical of the scientific establishment in the way Galois was treated; after all, his work had been in the hands of Cauchy, Fourier and Poisson, three giants of early nineteenth century mathematics and they all failed to appreciate the profound nature of Galois' ideas. There are certainly those who have tried to make Galois out as a misunderstood genius, surrounded by idiots, who, faced with repeated rejection, lost his way in life. The truth would appear to be somewhat different and it seems highly likely that Galois was the architect of much of what came his way. He clearly made little effort to appeal to those in positions of responsibility within the scientific community; Sophie Germain described him as clever, but impertinent, for example.

Nevertheless, it is without question that this angry and volatile young man developed mathematical ideas that were completely novel, at a time when there was no accepted language in which to describe them and it does seem appropriate that he is remembered primarily for the central role he played in the creation of abstract algebra.

6.1 Galois Groups

One of Galois' fundamental insights was that it is possible to associate with any polynomial a permutation group which describes the nature of its roots. For a polynomial of degree n, this group, now known as the Galois group of the polynomial, is a subgroup of the symmetric group S_n and it encodes information about how the roots can be found from the coefficients using the basic arithmetic operations $+, -, \times, \div$ and extracting square roots, cube roots etc. Given that there are only finitely many subgroups of S_n, this procedure partitions all polynomials of degree n into one of finitely many types and general observations can be made about all polynomials of a particular type. It is an incredibly powerful concept.

Before giving a description of Galois groups as they are usually thought of today, we will try and describe things more in the spirit of Galois' approach. To Galois, the group that now bears his name was the group of permutations of the roots of a polynomial that preserve all the polynomial relations between the roots. By this we mean that if g is a polynomial in the roots $x_1, \dots, x_n$ with coefficients in F, and $\sigma \in S_n$ is a permutation, then σ is in the Galois group if and only if

$$g(x_1, \dots, x_n) = 0 \quad \text{implies} \quad g(\sigma(x_1, \dots, x_n)) = 0$$

for all possible g. By adjoining new quantities to F, there are more possible polynomials g and the associated Galois group will shrink in size. Eventually, after we have added sufficient new quantities to the base field, the only element of S_n that will be left in the Galois group will be the identity permutation and the adjoined quantities will tell us about the solution to the original polynomial.

For concreteness, let's think of this in the context of quartic equations with rational coefficients. A quartic equation has four roots and there are 4!=24 ways of permuting them (equivalently, S_4 has 24 elements). We know that there are polynomials in the roots which take rational values, even if the roots are irrational. In particular, all the symmetric polynomials have this property. Thus for the polynomial $x^4+bx^3+cx^2+dx+e$ we know for example that $x_1+x_2+x_3+x_4 = -b$. If we apply any permutation in S_4 to this relation, it remains true. If the symmetric relationships such as this one are the only algebraic relations between the roots, with coefficients in $\mathbb{Q}$, then the Galois group is S_4.

If, on the other hand, there are additional algebraic relations between the roots, then this will put restrictions on which elements of S_4 are in the Galois group. As we enlarge the base field $\mathbb{Q}$, the number of possible polynomial relations g will increase and thus the Galois group will become more constrained. In Chap. 4 we saw that the roots of a quartic are related to the roots y_1, y_2, y_3 of an associated cubic via the relations

$$x_1x_2 + x_3x_4 = y_1, \quad x_1x_3 + x_2x_4 = y_2, \quad x_1x_4 + x_2x_3 = y_3$$

and so if any of y_1, y_2, y_3 are in $\mathbb{Q}$ we will have some additional relations between the roots that the Galois group needs to preserve. In this way we have sense of the way in which the Galois group encodes information about the roots of a polynomial.

Example 6.1 In Chap. 2 we considered the quartic $x^{+}x^{+} \in \mathbb{Q}[x]$ and used Ferrari's method to show that the roots are

$$x_{=}\sqrt{\ } - +\mathrm{i}\sqrt{\sqrt{-}}, \quad x_{=}\sqrt{\ } - -\mathrm{i}\sqrt{\sqrt{-1}},$$

$$x_3 = -\sqrt{2} - 1 + \sqrt{\sqrt{2} + 1}, \quad x_4 = -\sqrt{2} - 1 - \sqrt{\sqrt{2} + 1}.$$

We can then calculate

$$\begin{aligned} x_1x_2 + x_3x_4 = 4, \quad x_1x_3 + x_2x_4 &= -2 + 2\mathrm{i}, \\ x_1x_4 + x_2x_3 &= -2 - 2\mathrm{i}, \end{aligned} \tag{6.1}$$

where $4, -2 \pm 2\mathrm{i}$ are the three roots of $y^3 - 8y - 32$.

The relation $x_1x_2 + x_3x_4 = 4$ is not preserved by all permutations in S_4. If we apply the permutation (x_1x_3) we get $x_3x_2 + x_1x_4 = 4$, which is false. Hence the Galois group of $x^4 + 4x^3 + 2$ over $\mathbb{Q}$ is not S_4, but some subgroup of it. All the elements of S_4 are listed in Sect. 4.4 and it is a simple matter to check which ones do preserve the relation $x_1x_2 + x_3x_4 = 4$. They are

$$\begin{gathered} id, (x_1x_2), (x_3x_4), (x_1x_2)(x_3x_4), (x_1x_3)(x_2x_4), \\ (x_1x_4)(x_2x_3), (x_1x_3x_2x_4), (x_1x_4x_2x_3), \end{gathered}$$

which is $D_8^{(1)}$ (see (4.4)) and so only 8 of the 24 elements of S_4 remain.

If we adjoin i to the base field $\mathbb{Q}$, then we now have two additional relationships to preserve, involving $x_1x_3 + x_2x_4$ and $x_1x_4 + x_2x_3$. Each of these individually is preserved by a set of 8 permutations, but the intersection of all of these three sets of 8 is in fact just

$$V_4 = \{id, (x_1x_2)(x_3x_4), (x_1x_3)(x_2x_4), (x_1x_4)(x_2x_3)\},$$

(continued)

Example 6.1 (continued)

which preserves all three of the relations in (6.1). If we next adjoin $\sqrt{2}$, so that we are working over $\mathbb{Q}(\mathrm{i}, \sqrt{2})$, we find that we have new relations:

$$x_1 + x_2 = 2(\sqrt{2} - 1), \quad x_3 + x_4 = -2(\sqrt{2} + 1)$$

and now only *id* and $(x_1x_2)(x_3x_4)$ preserve all the relations between the roots and we note that $\{id, (x_1x_2)(x_3x_4)\} \cong C_2$. Finally, we adjoin $\alpha = \sqrt{\sqrt{2} - 1}$ and we have $x_1 - x_2 = 2\mathrm{i}\alpha$ and $x_3 - x_4 = 2/\alpha$ and the only permutation that preserves all the algebraic relations is the identity.

How is this related to the solution of the quartic? What this process shows is that, for this equation, we start with a group of 8 elements and then adjoin three quantities, each of which halves the size of the group. Dividing the size of the group by two is equivalent to taking a square root and hence the solution involves three square roots: $\sqrt{2}$, $\sqrt{\sqrt{2} - 1}$ and $\mathrm{i} = \sqrt{-1}$. ♦

Of course, in this case we started by knowing the answer, but even if we didn't know the solution we would still be able to infer a great deal about the nature of the solutions if we just knew the Galois group and how it can be broken down into smaller groups.

6.2 Automorphisms and Fixed Fields

A detailed description of the development of Galois theory from its early beginnings to its modern formulation in the twentieth century can be found in [33]. In order to understand what the Galois group looks like from a modern perspective, we need to introduce the concepts of an automorphism and that of its fixed field.

An **automorphism** of a field F is a map from F to itself which preserves the field structure. In a sense, it is a symmetry of F; just as a rotation of a square by 90° results in a geometric object that has exactly the same properties as the original square, applying an automorphism to a field F yields a field which has exactly the same algebraic properties as F. For example, the process of complex conjugation, which maps a complex number $z = x + \mathrm{i}y$ (x and y real) into its complex conjugate $\overline{z} = x - \mathrm{i}y$ is an automorphism of $\mathbb{C}$. If we represent complex numbers geometrically in the (x, y)-plane then complex conjugation is also seen to be just reflection in the real axis so we can visualise it as a geometrical symmetry as well. This automorphism is a manifestation of the fact that i and $-\mathrm{i}$ are algebraically indistinguishable over $\mathbb{R}$ (meaning that there is no polynomial in $\mathbb{R}[x]$ which has one as a root but not the other) and hence which one we label as i and which we label as $-\mathrm{i}$ is immaterial.

The first condition that any automorphism must satisfy is that it must set up a 1-1 correspondence between the elements of F and themselves. In other words an automorphism ϕ must be a **bijection**, which means first that $\phi(a) = \phi(b)$ if and only if $a = b$ and secondly that if $b \in F$, there is an $a \in F$ such that $\phi(a) = b$. Secondly, in order to preserve the structure of a field, which is defined in terms of the operations of arithmetic, we require that for all $a, b \in F$,

$$\phi(a \pm b) = \phi(a) \pm \phi(b), \quad \phi(ab) = \phi(a)\phi(b), \quad \phi(a/b) = \phi(a)/\phi(b),$$

excluding division by zero. Setting $a = b = 0$ in the first of these shows that $\phi(0) = 0$ for any automorphism. Since no other element of F can map to zero (as ϕ is a bijection) and $\phi(1 \times 1) = \phi(1) \times \phi(1)$, we must have $\phi(1) = 1$ and then it follows that $\phi(p/q) = \phi(p)/\phi(q) = p/q$ for any integers p and q. Thus any automorphism ϕ of a field F fixes all elements of $\mathbb{Q}$.

The set of all automorphisms of a field K, denoted $\text{Aut}(K)$, forms a group under composition. In other words, automorphisms satisfy the four group axioms of closure, associativity, identity and inverses, as described in Sect. 4.2. If F is a subfield of K, then the automorphisms of the field K that fix every element of F (so-called F-**automorphisms of** K) also form a group under composition denoted by $\text{Aut}(K/F)$ which is in fact a subgroup of $\text{Aut}\,K$. This is essentially just a statement of the fact that if you perform two automorphisms which both fix F one after the other, then the resulting automorphism also fixes F. Note that elements of $\text{Aut}(\mathbb{Q}(S)/\mathbb{Q})$ are simply the automorphisms of $\mathbb{Q}(S)$, since all automorphisms fix $\mathbb{Q}$; hence we will usually write $\text{Aut}(\mathbb{Q}(S))$ instead of $\text{Aut}(\mathbb{Q}(S)/\mathbb{Q})$.

Example 6.2 The automorphism of $\mathbb{C}$ given by $\phi(x + iy) = x - iy$, representing complex conjugation, is an element of $\text{Aut}(\mathbb{C}/\mathbb{R})$ since all elements of $\mathbb{R}$ are fixed by ϕ. ♦

Now let H be a subgroup of $\text{Aut}(K)$. Elements of H are automorphisms and so they fix elements of $\mathbb{Q}$. However, there may be other elements of the field K which are fixed by all the automorphisms in H. The **fixed field** L of H is defined as the subfield of K which is fixed by all the elements of H. In more technical language, $L = \{x \in K : \phi(x) = x \text{ for all } \phi \in H\}$. The following proposition is a rather technical exercise in linear algebra [13, Thm. 7.3.1(b)]:

Proposition 6.1 *Let $H \leq \text{Aut}(K)$ and L be its fixed field. Then the degree of the field extension K/L is given by $[K : L] = |H|$.*

Note that if $H = \{id\}$, so that all elements of K are fixed by H, we have $L = K$ and $[K : K] = 1$ as required.

Example 6.3 Take $K = \mathbb{C}$ and $H = \{id, \phi\}$, where ϕ is complex conjugation as in the previous example, for which $\phi^2 = id$ and so $H \cong C_2$. The fixed field of H is $\mathbb{R}$ and we have $|H| = 2 = [\mathbb{C} : \mathbb{R}]$; see Example 5.1. ♦

Let L be the fixed field of $\mathrm{Aut}(K/F)$. Since, by definition, any element of $\mathrm{Aut}(K/F)$ fixes F, we must have $F \subseteq L$, but they are not necessarily the same, as the next example illustrates.

Example 6.4 Consider $\mathbb{Q}(\sqrt[3]{2})$. We have, for any automorphism $\phi \in \mathrm{Aut}(\mathbb{Q}(\sqrt[3]{2}))$,

$$\phi(\sqrt[3]{2})^3 = \phi(2) = 2.$$

Since $\mathbb{Q}(\sqrt[3]{2}) \subset \mathbb{R}$ and there is only one real cube root of 2, it is apparent that $\phi(\sqrt[3]{2}) = \sqrt[3]{2}$ and hence that ϕ maps any number of the form $a+b\sqrt[3]{2}$ to itself. It follows that the only automorphism of $\mathbb{Q}(\sqrt[3]{2})$ is the identity automorphism. The fixed field of $\mathrm{Aut}(\mathbb{Q}(\sqrt[3]{2}))$ is therefore $\mathbb{Q}(\sqrt[3]{2})$ which is strictly larger than $\mathbb{Q}$. In this example we have, see Example 5.9,

$$|\mathrm{Aut}(\mathbb{Q}(\sqrt[3]{2}))| = 1, \quad [\mathbb{Q}(\sqrt[3]{2}) : \mathbb{Q}] = 3.$$ ♦

It is possible to develop a theory of automorphisms for arbitrary field extensions, but for our purposes we are only interested in the special situation when K is the splitting field of a separable polynomial $f \in F[x]$ (separable means no multiple roots). An extension K/F, where K is the splitting field of a separable polynomial $f \in F[x]$ is known as a **Galois extension** and in this case we write $\mathrm{Gal}(K/F)$ for the group of F-automorphisms of K, rather than $\mathrm{Aut}(K/F)$. The **Galois group** of a separable polynomial $f \in F[x]$, denoted $\mathrm{Gal}(f, F)$, is defined to be $\mathrm{Gal}(K/F)$, where K is the splitting field for f over F, i.e. $K = F(x_1, \ldots, x_n)$, where $x_1, \ldots, x_n$ are the roots of f. Where it is clear what the base field is, we will usually write G_f rather than $\mathrm{Gal}(f, F)$ for brevity. The word 'extension' is often dropped and K/F simply refered to as being Galois.

One of the most important properties of an element $\phi \in \mathrm{Gal}(f, F)$ relates to the effect of ϕ on the roots of f. If we write $f(x) = a_n x^n + a_{n-1}x^{n-1} + \cdots + a_1 x + a_0$, then, since $\phi(a_i) = a_i$ for $a_i \in F$, using the properties of automorphisms we can show that, for any α,

$$\phi(f(\alpha)) = f(\phi(\alpha)).$$

If we now let α be a root of f, so that $f(\alpha) = 0$, the left-hand side is $\phi(0)$ which is zero since ϕ is an automorphism. Hence, the right-hand side must also be zero and so $\phi(\alpha)$ must also be a root of f. Therefore ϕ maps any root of f to a root of f (it might be the same one) which means that ϕ permutes the roots of f amongst themselves.

Any element of the splitting field $K = F(x_1, \ldots, x_n)$ can be written as a rational function of the roots with coefficients in F; this is just from the definition of a field. If we apply an automorphism $\phi \in G_f$ to an element of K then, since ϕ fixes elements of F, it will be given by the same rational function of $\phi(x_1), \ldots, \phi(x_n)$. It follows that any automorphism $\phi \in G_f$ is completely determined by its effect on the roots of f and since elements of G_f permute the roots, we see that G_f can be interpreted as a group of permutations acting on the set of roots $\{x_1, \ldots, x_n\}$. This is an incredibly powerful idea and we will use these different interpretations of the Galois group interchangeably, with the same notation G_f, $\mathrm{Gal}(f, F)$ or $\mathrm{Gal}(K/F)$ used for the group of automorphisms of $K = F(x_1, \ldots, x_n)$ which fix F and the group of permutations of the roots $x_1, \ldots, x_n$ of f.

While any automorphism in G_f is equivalent to a permutation of the roots of f, not all possible permutations need correspond to automorphisms and so G_f will be equivalent to some subgroup of the symmetric group S_n. Since G_f is a subgroup of S_n and $|S_n| = n!$, Lagrange's theorem (see Sect. 4.4) shows that $|G_f|$ is a factor of $n!$. Unlike in Example 6.4, for a Galois extension we always have $|\mathrm{Gal}(K/F)| = [K : F]$ and that F is precisely the fixed field of $\mathrm{Gal}(K/F)$. These two results are part of what is often referred to as the **fundamental theorem of Galois Theory** and their proof is somewhat involved; see, for example, [13, §7.1]. In fact, it can be shown that:

Proposition 6.2 *The following three statements are equivalent:*

- *K is the splitting field of a separable polynomial $f \in F[x]$,*
- *F is precisely the fixed field of* $\mathrm{Gal}(K/F)$,
- $|\mathrm{Gal}(K/F)| = [K : F]$.

Example 6.5 The splitting field of $x^2 - 2$ over $\mathbb{Q}$ is $\mathbb{Q}(\sqrt{2})$ (see Example 5.10). The map $\phi : \mathbb{Q}(\sqrt{2}) \to \mathbb{Q}(\sqrt{2})$ defined by $\phi(a + b\sqrt{2}) = a - b\sqrt{2}$, $a, b \in \mathbb{Q}$, is an automorphism, as is easily checked. Clearly ϕ sends $\sqrt{2}$, which is a root of $x^2 - 2$ to the other root, $-\sqrt{2}$. The identity automorphism maps $\sqrt{2}$ to itself and since there are no other roots to $x^2 - 2$, this completes the set of automorphisms. We can equally well think of ϕ as a permutation in S_2, namely (x_1x_2), and clearly ϕ^2 is the identity. Hence the group $\text{Gal}(\mathbb{Q}(\sqrt{2})) = \{id, \phi\}$, which is isomorphic to C_2, the cyclic group of order 2. In this example we have (see Example 5.3)

$$|\,\text{Gal}(\mathbb{Q}(\sqrt{2}))| = [\mathbb{Q}(\sqrt{2}) : \mathbb{Q}] = 2.$$

Also, the fixed field of $\text{Gal}(\mathbb{Q}(\sqrt{2}))$ is precisely $\mathbb{Q}$, since $\phi(\alpha) = \alpha$ only if $b = 0$, in which case $\alpha \in \mathbb{Q}$. ♦

Example 6.6 Building on Examples 5.9 and 5.11. The number $\sqrt[3]{2}$ is one root of the polynomial $x^3 - 2$, which is irreducible over $\mathbb{Q}$ (and hence separable). But $\mathbb{Q}(\sqrt[3]{2}) \subset \mathbb{R}$ is clearly not the splitting field for $x^3 - 2$ since the other two roots are complex. Hence $\mathbb{Q}(\sqrt[3]{2})/\mathbb{Q}$ is not a Galois extension. This is consistent with our previous observations (in Example 6.4) that the fixed field of $\text{Aut}(\mathbb{Q}(\sqrt[3]{2}))$ is strictly larger than $\mathbb{Q}$ and $|\,\text{Aut}(\mathbb{Q}(\sqrt[3]{2}))| \neq [\mathbb{Q}(\sqrt[3]{2}) : \mathbb{Q}]$. Similarly, neither $\mathbb{Q}(\omega\sqrt[3]{2})/\mathbb{Q}$ or $\mathbb{Q}(\omega^2\sqrt[3]{2})/\mathbb{Q}$ are Galois extensions.

However, $\mathbb{Q}(\sqrt[3]{2}, \omega)$ is the splitting field for $x^3 - 2$ over $\mathbb{Q}$ and $\mathbb{Q}(\omega)$ is the splitting field for the irreducible polynomial $x^2 + x + 1$ and hence both are Galois extensions. ♦

6.3 Galois Groups of Irreducible Polynomials

Any polynomial can be written as a product of irreducible factors in an essentially unique way (see Sect. 3.2). We can think of these factors as being independent from each other and it turns out that the Galois group interacts with each factor separately. Thus G_f permutes the roots of each irreducible factor amongst themselves and there is no permutation that takes a root of one factor to the root of a different factor.

This means that if $f \in F[x]$ is irreducible, then, and only then, G_f is a **transitive** subgroup of S_n, a result which is due to Camille Jordan in 1870, though appears to have been understood by Galois. What this means is that there is a permutation in G_f that takes each root to any of the other roots or, in other words, if x_i and x_j are any two roots, then there is a permutation σ in the Galois group such that $\sigma(x_i) = x_j$. If f is reducible, then G_f acts transitively on the roots of each of the irreducible factors of f. Note that transitivity is a property of how a group acts on the set $\{x_1, \ldots, x_n\}$. It is possible to have two isomorphic subgroups of S_n, such that one is transitive and the other is not.

Example 6.7 In Chap. 4 we identified four subgroups of S_4 of order 4. They were

$$V_4 = \{id, (x_1x_2)(x_3x_4), (x_1x_3)(x_2x_4), (x_1x_4)(x_2x_3)\},$$

and three others, one of which is

$$\{id, (x_1x_2), (x_3x_4), (x_1x_2)(x_3x_4)\}.$$

By inspection, we see that V_4 acts transitively on the roots x_1, x_2, x_3, x_4 (if you take any two roots, x_i and x_j, there is a permutation in V_4 that takes x_i to x_j) but none of the others do. In the second example above, there is no permutation which takes x_1 to x_3, for example. ♦

It follows from Proposition 5.4 that if $f \in F[x]$ is irreducible over F, then $|G_f|$ is a multiple of the degree of f. Thus, in determining the Galois group of an irreducible polynomial $f \in F[x]$ of degree n, viewed as a permutation of the roots $\{x_1, \ldots, x_n\}$, we only need to consider transitive subgroups of S_n, whose order is divisible by n. In fact, it can be shown that any transitive subgroup of S_n will necessarily have an order that is a multiple of n; see [13, Ex. 6, §6.3].

We showed in Sect. 4.3 that the square root of the discriminant, δ, is fixed by even permutations and δ maps to $-\delta$ for odd permutations. Thus provided $\delta \neq 0$ we see that if δ is fixed by the Galois group, that group can only contain even permutations. Hence:

Proposition 6.3 *If the discriminant Δ of a separable polynomial $f \in F[x]$ of degree n is a square in F, i.e. $\delta \neq 0$ and $\delta = \sqrt{\Delta} \in F$, the Galois group must be a subgroup of A_n, and conversely.*

Example 6.8 Consider

$$f = x^3 + bx^2 - (b+3)x + 1.$$

The rational root test shows that this is irreducible unless $b = -3/2$ (and then $f = (x+1)(x-2)(x-1/2)$). The discriminant of f is

$$\Delta = (b^2 + 3b + 9)^2$$

and so G_f is a subgroup of A_3 (i.e. either A_3 or $\{id\}$). Unless $b = -3/2$, this must be A_3 as its order must be a multiple of 3 and when $b = -3/2$ it is the identity since all the roots are in $\mathbb{Q}$. ♦

We will now go through a few examples in some detail to illustrate the many ideas that have been introduced in this chapter. In due course, we will be able to take some short cuts to get to the Galois group, but these examples are helpful in showing how the theory fits together.

Example 6.9 Let $f \in \mathbb{Q}[x]$ be given by

$$f(x) = x^4 - 5x^2 + 6 = (x^2 - 2)(x^2 - 3)$$

which is separable and reducible. The roots are $\pm\sqrt{2}$ and $\pm\sqrt{3}$ and so the splitting field is $\mathbb{Q}(\sqrt{2}, \sqrt{3})$ and $G_f = \mathrm{Gal}(\mathbb{Q}(\sqrt{2}, \sqrt{3}))$. The elements of G_f are automorphisms of $\mathbb{Q}(\sqrt{2}, \sqrt{3})$ and these are completely determined by their effect on $\sqrt{2}$ and $\sqrt{3}$. Now

$$\phi(\sqrt{2})^2 = \phi(2) = 2$$

and so $\phi(\sqrt{2}) = \pm\sqrt{2}$ and similarly for $\sqrt{3}$. There are thus four possibilities for ϕ, which we can label $\phi_1, \ldots, \phi_4$, with ϕ_1 being the identity automorphism. The others are defined by their actions on $\sqrt{2}$ and $\sqrt{3}$:

$$\phi_2 : \sqrt{2} \mapsto -\sqrt{2},\ \sqrt{3} \mapsto \sqrt{3},$$
$$\phi_3 : \sqrt{2} \mapsto \sqrt{2},\ \sqrt{3} \mapsto -\sqrt{3},$$
$$\phi_4 : \sqrt{2} \mapsto -\sqrt{2},\ \sqrt{3} \mapsto -\sqrt{3}.$$

(continued)

Example 6.9 (continued)

In terms of the permutations of the roots, if we label them

$$x_1 = \sqrt{2}, \quad x_2 = -\sqrt{2}, \quad x_3 = \sqrt{3}, \quad x_4 = -\sqrt{3}, \tag{6.2}$$

we see that (using the same notation to represent automorphisms and their equivalent permutations)

$$\phi_2 = (x_1 x_2), \quad \phi_3 = (x_3 x_4), \quad \phi_4 = (x_1 x_2)(x_3 x_4). \tag{6.3}$$

We can see by inspection that the roots satisfy the non-symmetric relations $x_1 + x_2 = x_3 + x_4 = 0$, $x_1 x_2 = -2$ and $x_3 x_4 = -3$. Hence the elements of the Galois group must each preserve all of these relations, as those in (6.3) clearly do. The four elements of the group are $\{id, \sigma, \tau, \sigma\tau\}$, where $\sigma = \phi_2$ and $\tau = \phi_3$, with $\sigma^2 = \tau^2 = id$ and $\sigma\tau = \tau\sigma$. Hence G_f is isomorphic to the Klein 4-group, V_4. Which of the three possibilities listed in (4.5) is the Galois group depends on how we label the roots and with the choice in (6.2), we have the first of the three, but if we had chosen the order so that $x_1 = -x_3$, we would have the second in the list and with $x_1 = -x_4$, the third.

A simple calculation shows that the discriminant $\Delta = 96$, which is not a square in $\mathbb{Q}$ and so the Galois group is not contained in A_4. This is also evident from the presence of odd permutations in (6.3). Since f is reducible, the group represented by (6.3) is not transitive, which can be seen since, for example, there is no permutation in the group which takes x_1 to x_3. However, the group acts transitively on $\{x_1, x_2\}$ and $\{x_3, x_4\}$, corresponding to the irreducible factors of f.

We also know from Example 5.4 that $\mathbb{Q}(\sqrt{2}, \sqrt{3}) = \mathbb{Q}(\sqrt{2} + \sqrt{3})$ and so equally we have $G_f = \text{Gal}(\mathbb{Q}(\sqrt{2}+\sqrt{3}))$ and automorphisms of $\mathbb{Q}(\sqrt{2}+\sqrt{3})$ are determined by their effect on $\sqrt{2} + \sqrt{3}$. Any such automorphism satisfies

$$\phi(\sqrt{2} + \sqrt{3}) = \phi(\sqrt{2}) + \phi(\sqrt{3}) = \pm\sqrt{2} \pm \sqrt{3}$$

with four possible values. Hence, apart from the identity ϕ_1, we have

$$\begin{aligned} \phi_2(\sqrt{2} + \sqrt{3}) &= -\sqrt{2} + \sqrt{3}, \\ \phi_3(\sqrt{2} + \sqrt{3}) &= \sqrt{2} - \sqrt{3}, \\ \phi_4(\sqrt{2} + \sqrt{3}) &= -\sqrt{2} - \sqrt{3}. \end{aligned}$$

(continued)

Example 6.9 (continued)

In Example 5.6 we showed that the minimal polynomial for $\sqrt{2}+\sqrt{3}$ over $\mathbb{Q}$ is

$$g(x) = x^4 - 10x^2 + 1$$

with roots

$$x_1 = \sqrt{2}+\sqrt{3}, \quad x_2 = -\sqrt{2}+\sqrt{3},$$
$$x_3 = \sqrt{2}-\sqrt{3}, \quad x_4 = -\sqrt{2}-\sqrt{3}$$

and we can work out the effect of ϕ_2, ϕ_3, ϕ_4 on x_2, x_3 and x_4 with the aid of

$$x_1x_2 = x_3x_4 = 1, \quad x_3 = -x_2, \quad \text{and} \quad x_4 = -x_1. \tag{6.4}$$

In terms of the permutations of these roots we find

$$\phi_2 = (x_1x_2)(x_3x_4), \quad \phi_3 = (x_1x_3)(x_2x_4), \quad \phi_4 = (x_1x_4)(x_2x_3)$$

all of which clearly preserve the relations in (6.4). Note that the discriminant of g is $\Delta = 147{,}456 = 384^2$ and so the Galois group must be a subgroup of A_4. In this case the polynomial is irreducible and hence the group must also be transitive, as is easily checked. Again $\phi_2^2 = \phi_3^2 = id$ and $\phi_4 = \phi_2\phi_3 = \phi_3\phi_2$ and we see that in this case G_g is V_4, rather than simply being isomorphic to it. Hence

$$G_f = \text{Gal}(x^4 - 5x^2 + 6, \mathbb{Q}) \cong G_g = \text{Gal}(x^4 - 10x^2 + 1, \mathbb{Q}) = V_4$$

and G_g acts transitively on the roots while G_f does not. ♦

Example 6.10 Let $f \in \mathbb{Q}[x]$ be given by

$$f(x) = x^4 - 4x^2 + 2.$$

This is a quadratic in x^2 and the roots are easily found using the quadratic formula:

$$x_1 = \sqrt{2+\sqrt{2}}, \quad x_2 = \sqrt{2-\sqrt{2}},$$
$$x_3 = -\sqrt{2+\sqrt{2}}, \quad x_4 = -\sqrt{2-\sqrt{2}}.$$

(continued)

Example 6.10 (continued)

If we write $\alpha = \sqrt{2+\sqrt{2}}$, then we have

$$\begin{gathered} x_1 = \alpha, \quad x_2 = (\alpha^2 - 2)/\alpha, \quad x_2^2 = 4 - \alpha^2, \\ x_3 = -x_1, \quad x_4 = -x_2. \end{gathered} \tag{6.5}$$

We showed in Example 5.15 that $\mathbb{Q}(\alpha)$ is the splitting field for f over $\mathbb{Q}$ and that $[\mathbb{Q}(\alpha) : \mathbb{Q}] = 4$. To determine the group structure we just need to look at the effect of any automorphism on x_1 and since x_1 must map to one of the roots, there are four possibilities. Let $\phi_j(x_1) = x_j$, $j = 1, \ldots, 4$ and then we can use (6.5) to determine the effect of these automorphisms on the other roots. Clearly ϕ_1 is the identity and we have

$$\phi_2(x_2) = \phi_2\left(\frac{\alpha^2 - 2}{\alpha}\right) = \frac{\phi_2(\alpha)^2 - 2}{\phi_2(\alpha)} = \frac{x_2^2 - 2}{x_2} = -\alpha = x_3,$$

$$\phi_2(x_3) = -\phi_2(x_1) = -x_2 = x_4,$$

$$\phi_2(x_4) = -\phi_2(x_2) = -x_3 = x_1.$$

Thus ϕ_2 is the permutation $(x_1x_2x_3x_4)$, which is an odd permutation and so we know immediately that the Galois group is not a subgroup of A_4 (this also follows from the fact that $\Delta = 2048$ is not a square in $\mathbb{Q}$). In a similar way we can show that ϕ_3 is the permutation $(x_1x_3)(x_2x_4)$ and ϕ_4 is the permutation $(x_1x_4x_3x_2)$. If we write $\phi_2 = \sigma$ then $\phi_3 = \sigma^2$ and $\phi_4 = \sigma^3$ and so the group has the elements $\{id, \sigma, \sigma^2, \sigma^3\}$ with $\sigma^4 = id$ and hence is cyclic:

$$\mathrm{Gal}(x^4 - 4x^2 + 2, \mathbb{Q}) \cong C_4.$$ ♦

Example 6.11 Let $f = x^4 - 2 \in \mathbb{Q}[x]$. We saw in Example 5.13 that the splitting field for f is $\mathbb{Q}(\sqrt[4]{2}, \mathrm{i})$ and that $[\mathbb{Q}(\sqrt[4]{2}, \mathrm{i}) : \mathbb{Q}] = 8$. Write $\alpha = \sqrt[4]{2}$. The roots of f are

$$x_1 = \alpha, \quad x_2 = -\alpha, \quad x_3 = \mathrm{i}\alpha, \quad x_4 = -\mathrm{i}\alpha$$

and we can calculate $\Delta = -2048$, which is not a square in $\mathbb{Q}$ and so the Galois group is not a subgroup of A_4. For any automorphism $\phi \in G_f$ we know that $\phi(\alpha)$ will take one of the four values $\pm\alpha$, $\pm\mathrm{i}\alpha$. Furthermore $\phi(\mathrm{i})^2 = \phi(-1) = -1$ and so $\phi(\mathrm{i}) = \pm\mathrm{i}$. Thus we have 8 possibilities. Let

(continued)

Example 6.11 (continued)
$\phi_j(\mathrm{i}) = \mathrm{i}$ for $j = 1, \dots, 4$ and $\phi_j(\mathrm{i}) = -\mathrm{i}$ for $j = 5, \dots, 8$, and

$$\phi_1(\alpha) = \phi_5(\alpha) = \alpha, \qquad \phi_2(\alpha) = \phi_6(\alpha) = \mathrm{i}\alpha,$$
$$\phi_3(\alpha) = \phi_7(\alpha) = -\alpha, \qquad \phi_4(\alpha) = \phi_8(\alpha) = -\mathrm{i}\alpha.$$

Then ϕ_1 is the identity and we can compute

$$\phi_2 = (x_1x_3x_2x_4), \quad \phi_3 = (x_1x_2)(x_3x_4), \quad \phi_4 = (x_1x_4x_2x_3),$$
$$\phi_5 = (x_3x_4), \quad \phi_6 = (x_1x_3)(x_2x_4),$$
$$\phi_7 = (x_1x_2), \quad \phi_8 = (x_1x_4)(x_2x_3).$$

If we write $\phi_2 = \sigma = (x_1x_3x_2x_4)$ and $\phi_7 = \tau = (x_1x_2)$, then the group elements can be seen to be

$$\{id, \sigma, \sigma^2, \sigma^3, \tau, \tau\sigma, \tau\sigma^2, \tau\sigma^3\},$$

with $\sigma^4 = \tau^2 = id$ and $\sigma\tau = \tau\sigma^3$. This group is the dihedral group $D_8^{(1)}$ once more. Thus

$$\mathrm{Gal}(x^4 - 2, \mathbb{Q}) \cong D_8. \qquad \blacklozenge$$

6.4 The Galois Correspondence

We now come to a central result in the modern presentation of Galois theory, though one that was anticipated to a large extent by Lagrange, over 50 years before Galois. We take as our starting point a polynomial with coefficients in a field F, i.e. $f \in F[x]$. If we adjoin the n roots $x_1, \dots, x_n$ to the field F we obtain the splitting field K and we have seen that, associated with K is the Galois group $G_f = \mathrm{Gal}(K/F)$ of the polynomial which can be thought of as a subgroup of the symmetric group S_n.

We can set up a correspondence between intermediate fields L ($F \subseteq L \subseteq K$) and subgroups H of $\mathrm{Gal}(K/F)$ as follows. Since K is the splitting field of a polynomial with coefficients in F, we can consider those coefficients to be part of the larger field L and then K will still be the splitting field and so K/L is also a Galois extension. For a given L we associate the subgroup $H = \mathrm{Gal}(K/L)$ and for a given subgroup H, we associate its fixed field. The fundamental theorem of Galois theory asserts that this correspondence between all possible intermediate fields L and all possible subgroups H is a bijection, called the **Galois correspondence**. To every L, there corresponds a unique H and vice versa. Under this bijection L maps to the subgroup of G_f that fixes L and in the other direction the subgroup $H = \mathrm{Gal}(K/L)$ maps to its fixed field. As special cases, F corresponds to G_f and K corresponds to the trivial group $\{id\}$. We can illustrate the Galois correspondence by means of a diagram:

$$\begin{array}{ccccc} F & \subseteq & L & \subseteq & K \\ \updownarrow & & \updownarrow & & \updownarrow \\ G_f & \geq & H & \trianglerighteq & \{id\} \end{array}$$

First, note the inclusion reversing nature of this correspondence. As the field is enlarged, the associated group becomes smaller. Second, we have indicated in the diagram that the trivial group $\{id\}$ is a normal subgroup of any group and this is because normal subgroups correspond to Galois extensions and we have seen above that K/L is always Galois. Thus, as well as the correspondence between intermediate fields of the splitting field and subgroups of the Galois group, the fundamental theorem asserts that L is a Galois extension of F if and only if $H \trianglelefteq G_f$. In that case, $\mathrm{Gal}(L/F)$ is isomorphic to the quotient group G_f/H and hence

$$|\mathrm{Gal}(L/F)| = |G_f|/|H|.$$

We also have, see Proposition 6.1, that $|H| = [K : L]$ and hence

$$|G_f| = [K : F] = [K : L]\,[L : F].$$

The power of the fundamental theorem stems from the fact that it can be used to translate questions about intermediate fields into questions about finite groups, about which a great deal is known. One obvious consequence is that there are only finitely many intermediate fields in a Galois extension. The theorem allows us to explore the intermediate fields between the base field of a polynomial and its splitting field by considering the subgroups of the polynomial's Galois group; and the intermediate fields are key to understanding the nature of the roots of the polynomial.

Example 6.12 Let $f \in \mathbb{Q}[x]$ be given by $f = x^3 - 2$. The splitting field of f is $\mathbb{Q}(\sqrt[3]{2}, \omega)$ and so $G_f = \mathrm{Gal}(\mathbb{Q}(\sqrt[3]{2}, \omega))$. We have previously established (see Example 5.9) that $[\mathbb{Q}(\sqrt[3]{2}, \omega) : \mathbb{Q}] = 6$ and hence

$$|G_f| = [\mathbb{Q}(\sqrt[3]{2}, \omega) : \mathbb{Q}] = 6.$$

An element $\phi \in G_f$ is determined by its effect on the roots of $x^3 - 2$, namely

$$x_1 = \sqrt[3]{2}, \quad x_2 = \omega\sqrt[3]{2}, \quad x_3 = \omega^2\sqrt[3]{2}.$$

(continued)

Example 6.12 (continued)

The 6 elements of G_f give the 6 possible permutations of the three roots. We can take ϕ_1 as the identity, and

$$\phi_2 = (x_1x_2x_3), \quad \phi_3 = (x_1x_3x_2),$$
$$\phi_4 = (x_2x_3), \quad \phi_5 = (x_1x_3), \quad \phi_6 = (x_1x_2).$$

As it represents the permutations of three elements, $G_f = S_3$. The 6 elements of the group are $\{id, \sigma, \sigma^2, \tau, \sigma\tau, \sigma^2\tau\}$, where $\sigma = \phi_2$ and $\tau = \phi_4$, with $\sigma^3 = \tau^2 = id$ and $\tau\sigma = \sigma^2\tau$.

There are four proper subgroups of S_3: the alternating group $A_3 = \{id, \sigma, \sigma^2\} \cong C_3$ and the three groups $H_4 = \{id, \tau\}$, $H_5 = \{id, \sigma\tau\}$, $H_6 = \{id, \tau\sigma\}$, each of which is isomorphic to C_2. Note that $\tau = (x_2x_3)$ fixes x_1, whereas $\sigma\tau$ fixes x_2 and $\tau\sigma$ fixes x_3.

The Galois correspondence is illustrated below:

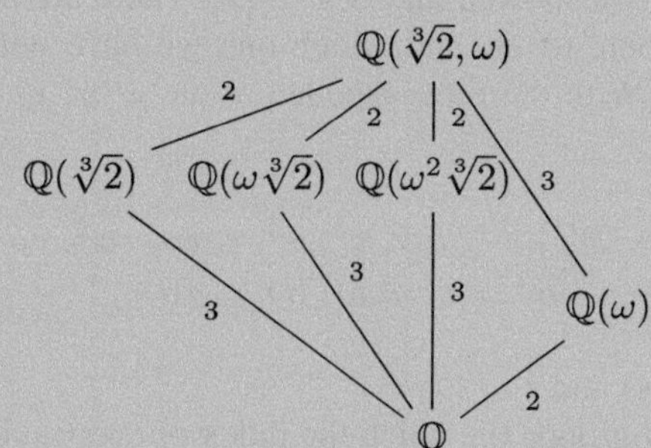

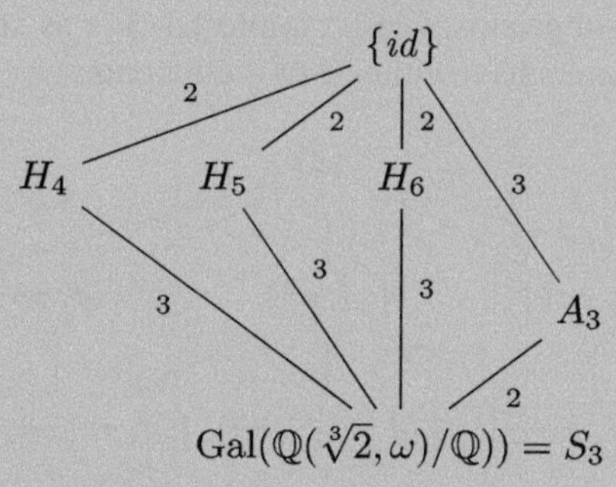

The fields in the upper diagram are the fixed fields of the corresponding groups below. In the top figure, reading downwards, the solid lines connect fields with their subfields and the numbers indicate the degree of the extension in each case. In the lower figure, reading upwards, the solid lines connect groups with their subgroups and the numbers indicate the ratio of the order of the group to that of its subgroup (i.e. the index of the subgroup).

(continued)

Example 6.12 (continued)

The trivial group, $\{id\}$, is a normal subgroup of all the other groups in the lower diagram, which corresponds to the fact that $\mathbb{Q}(\sqrt[3]{2}, \omega)$ is a Galois extension of all the other fields in the upper diagram. Note that A_3 is the only proper normal subgroup of S_3, which corresponds to the fact that $\mathbb{Q}(\omega)$ is the splitting field of $x^2 + x + 1$, whereas $\mathbb{Q}(\sqrt[3]{2})$, $\mathbb{Q}(\omega\sqrt[3]{2})$ and $\mathbb{Q}(\omega^2\sqrt[3]{2})$ are not splitting fields. ♦

Example 6.13 Let $f \in \mathbb{Q}[x]$ be given by $f = x^4 - 2$. We have shown in Example 5.13 that the splitting field of f is $\mathbb{Q}(\alpha, \mathrm{i})$, where $\alpha = \sqrt[4]{2}$, and that $[\mathbb{Q}(\alpha, \mathrm{i}) : \mathbb{Q}] = 8$. A basis for $[\mathbb{Q}(\alpha, \mathrm{i}) : \mathbb{Q}]$ is $\{1, \alpha, \alpha^2, \alpha^3, \mathrm{i}, \mathrm{i}\alpha, \mathrm{i}\alpha^2, \mathrm{i}\alpha^3\}$.

In Example 6.11 we showed that $G_f \cong D_8$. There are three subgroups of S_4 that are isomorphic to D_8 and which one we have will depend on how we label the roots. With the roots labelled as $x_1 = \alpha$, $x_2 = -\alpha$, $x_3 = \mathrm{i}\alpha$, $x_4 = -\mathrm{i}\alpha$, we have

$$G_f = D_8^{(1)} = \{id, \sigma, \sigma^2, \sigma^3, \tau, \tau\sigma, \tau\sigma^2, \tau\sigma^3\},$$
$$\sigma^4 = \tau^2 = id, \ \sigma\tau = \tau\sigma^3,$$

with $\sigma = (x_1x_3x_2x_4)$ and $\tau = (x_1x_2)$.

The lattice of subgroups (in which the index of every subgroup is 2 and so they are all normal subgroups; see Example 4.11) is as shown below, noting that all proper subgroups have either 2 or 4 elements:

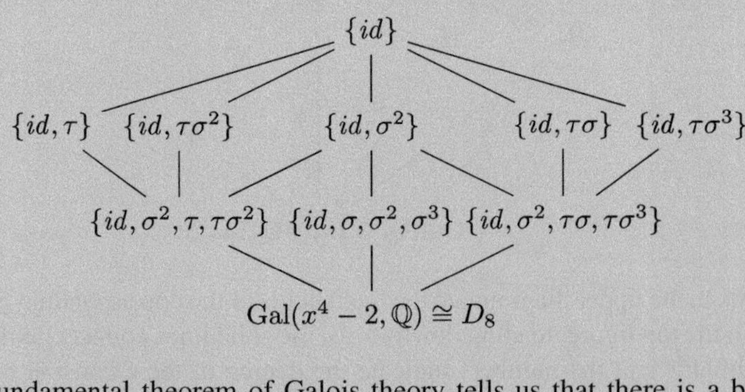

The fundamental theorem of Galois theory tells us that there is a bijection between the subgroups of $D_8^{(1)}$ and the intermediate fields between $\mathbb{Q}$ and $\mathbb{Q}(\alpha, \mathrm{i})$, each field being the fixed field of the corresponding subgroup. The subgroups are described by their effect on α and i and the degree of every extension is 2. For example $\{id, \sigma, \sigma^2, \sigma^3\}$ is generated by σ, which fixes i.

(continued)

Example 6.13 (continued)

Hence its fixed field is $\mathbb{Q}(\mathrm{i})$. Similarly, $\{id, \sigma^2, \tau, \tau\sigma^2\}$ is generated by σ^2 and τ, both of which fix α^2; hence its fixed field is $\mathbb{Q}(\alpha^2)$. Slightly less obviously, $\tau\sigma$ sends α to $-\mathrm{i}\alpha$ and i to $-\mathrm{i}$; hence it fixes $(1-\mathrm{i})\alpha$. Hence the fixed field of $\{id, \tau\sigma\}$ is $\mathbb{Q}((1-\mathrm{i})\alpha)$. Completing the diagram yields:

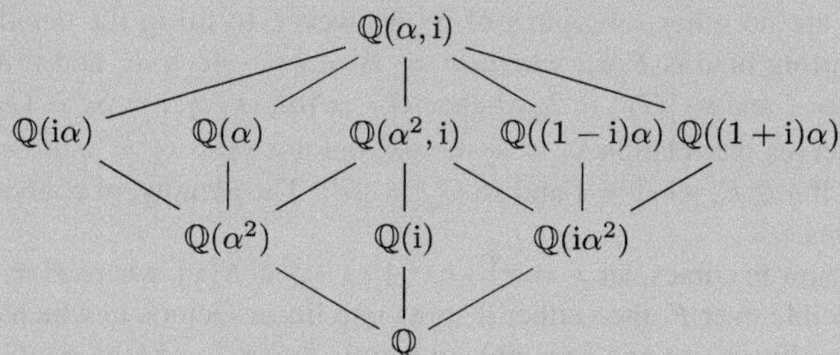

Reading up from the bottom, noting that $(1 \pm \mathrm{i})^2 = \pm 2\mathrm{i}$, each connection corresponds to the introduction of an additional square root. ♦

6.5 Galois Theory for Quadratics, Cubics and Quartics

After spending considerable time developing and illustrating the ideas that underpin Galois theory as it applies to polynomials in general, we will now return to quadratics, cubics and quartics and see how they fit into this framework. For ease of reference, we will collect together some key properties of Galois groups. Let $f \in F[x]$ be separable, with degree n, splitting field K and discriminant $\Delta = \delta^2$. Then

Proposition 6.4

P1: $|G_f| = [K : F]$ *is a factor of* $n!$,

P2: If f *is irreducible, then* G_f *is a transitive group acting on* $\{x_1, \ldots, x_n\}$ *and* $|G_f|$ *is a multiple of* n,

P3: G_f *is a subgroup of* A_n *if and only if* $\delta = \sqrt{\Delta} \in F$,

P4: The fixed field of G_f *is* F.

For quadratics $f = x^2 + bx + c \in F[x]$ with roots x_1 and x_2, there is not much to add to the content of Example 5.12. Note that property P3 actually tells us everything we need to know:

$$G_f = \begin{cases} A_2 = \{id\} & \text{if } \delta \in F \\ S_2 \cong C_2 & \text{if } \delta \notin F \end{cases},$$

since there are no other subgroups of S_2. However, to fill in the details, we note that the splitting field is $F(\delta)$, where $\delta^2 = \Delta = b^2 - 4c \in F$, and if $\delta \notin F$ then $[F(\delta) : F] = 2$ and so $|G_f| = 2$, whence $G_f = \{id, (x_1x_2)\} = S_2 \cong C_2$. Note that (x_1x_2) preserves the relation $x_1 + x_2 = -b$, but not $x_1 - x_2 = \sqrt{b^2 - 4c}$. On the other hand, if $\delta \in F$, $|G_f| = 1$ and so $G_f = \{id\}$. The identity, of course, preserves both relations.

Turning now to cubics, let $f = x^3+bx^2+cx+d \in F[x]$, where $F = \mathbb{Q}(b, c, d)$. If f is reducible over F, then either it splits into linear factors, in which case $G_f = \{id\}$, or into a linear and an irreducible quadratic factor, in which case $G_f \cong C_2$.

Now assume that f is irreducible over F (and hence $\Delta \neq 0$), which for cubic equations is equivalent to assuming that f has no roots in F. The roots of f are x_1, x_2, x_3 and we know that $x_1 + x_2 + x_3 = -b$. Hence the splitting field is $K = F(x_1, x_2, x_3) = F(x_1, x_2)$. We know that G_f is a subgroup of S_3 and hence $|G_f|$ $(= [K : F])$ is a factor of $|S_3|$ $(= 6)$. Since f is irreducible, each of its roots has minimal polynomial of degree 3 over F and so, in particular, $[F(x_1) : F] = 3$. Also, from Proposition 5.1, we know that $[K : F] = [K : F(x_1)][F(x_1) : F]$ and so $|G_f|$ must be either 3 or 6; it cannot be 1 or 2 (these cases correspond to reducible cubics as described in the previous paragraph). Hence we can conclude that $G_f = A_3$ or S_3 and, as we have seen earlier, this will depend on whether $\delta = \sqrt{\Delta} \in F$ or not. Thus

$$G_f = \begin{cases} A_3 \cong C_3 & \text{if } \delta \in F \\ S_3 & \text{if } \delta \notin F \end{cases}. \tag{6.6}$$

If $\delta \notin F$, then there is an intermediate field $L = F(\delta)$ such that $F \subset L \subset K$ with $[L : F] = 2$ and $[K : L] = 3$. Whether $\delta \in F$ or not, the splitting field is always obtained if we adjoin the square root of the discriminant and any one of the roots to the base field.

Example 6.14 Let

$$f = x^3 - 3x + 1 \in \mathbb{Q}[x],$$

which is a special case of Example 6.8. Irreducibility can be established using the rational root test. We can compute $\Delta = 81$ and so $\delta \in \mathbb{Q}$ in this case and

(continued)

Example 6.14 (continued)
$G_f = A_3$. Now let

$$f = x^3 - 4x + 1 \in \mathbb{Q}[x].$$

Irreducibility is established in the same way. We can compute $\Delta = 229$ and so $\delta \notin \mathbb{Q}$ in this case. Hence $G_f = S_3$.

In fact, it can be shown that the only polynomial of the form $x^3 - mx + 1$, where m is an integer, for which $G_f = A_3$ is $x^3 - 3x + 1$. Note that for $f = x^3 - 2x + 1$, we have $\Delta = 5$, but it is not the case that $G_f = S_3$ and this is because $f = (x-1)(x^2 + x - 1)$ is reducible. ♦

Example 6.15 The vast majority of cubics that are irreducible over $\mathbb{Q}$ have S_3 as their Galois group, since it is much less likely that the discriminant is a square in $\mathbb{Q}$ than otherwise. An example of an infinite family of cubics that have Galois group A_3 over $\mathbb{Q}$ is given in Example 6.8 and another is

$$x^3 - ax + a, \quad a = m^2 + m + 7, \quad m \in \mathbb{Z}.$$

Such cubics can be shown to be irreducible and the discriminant in this case is

$$(2m+1)^2(m^2 + m + 7)^2.$$ ♦

We can also derive Cardano's formula from knowledge about the Galois group. The roots of the cubic $x^3 + bx^2 + cx + d$ and its depressed form differ by the constant $b/3$. So if $F = \mathbb{Q}(b, c, d)$, both cubics have the same splitting field over F and therefore they have the same Galois group. We can therefore work with the depressed cubic $x^3 + px + q$, which we will assume is irreducible. If $p = 0$, we have $x^3 = -q$ and the three roots are the cube roots of $-q$. We now assume that $p \neq 0$. In the calculations that follow, we will make use of the following results concerning symmetric polynomials in the roots (the first three are the elementary symmetric polynomials that we have seen before, the rest can be derived from them as in Examples 4.5 and 4.6):

$$\begin{gathered} x_1 + x_2 + x_3 = 0, \quad x_1x_2 + x_1x_3 + x_2x_3 = p, \quad x_1x_2x_3 = -q, \\ x_1^2 + x_2^2 + x_3^2 = -2p, \quad x_1^3 + x_2^3 + x_3^3 = -3q, \\ x_1^2x_2 + x_1x_2^2 + x_1^2x_3 + x_1x_3^2 + x_2^2x_3 + x_2x_3^2 = 3q, \\ x_1^3x_2^3 + x_1^3x_3^3 + x_2^3x_3^3 = 3q^2 + p^3. \end{gathered} \tag{6.7}$$

Looking back at Examples 5.9 and 6.12 we see that the splitting field of the polynomial $x^3 - a$ over $F = \mathbb{Q}(a)$ will contain the cube root of unity ω, thus in order to be able to extract cube roots we need to adjoin ω to F, so in fact we take the base field to be $F = \mathbb{Q}(p, q, \omega)$. Let $L = F(\delta)$ so that $[L : F]$ is either 2 (if $\delta \notin F$) or 1 (if $\delta \in F$). From (6.6) and the remarks immediately following, we deduce that whether δ is in F or not, we have $\mathrm{Gal}(K/L) = A_3$, which is cyclic, and $[K : L] = 3$. Hence $\mathrm{Gal}(K/L) = \{id, \sigma, \sigma^2\}$ for some $\sigma \in A_3$ with $\sigma^3 = id$. We can take σ to be the permutation $(x_1x_2x_3)$ which is even and so fixes δ.

Now define u, v via

$$u = \tfrac{1}{3}\left(x_1 + \omega^2 x_2 + \omega x_3\right), \quad v = \tfrac{1}{3}\left(x_1 + \omega x_2 + \omega^2 x_3\right). \tag{6.8}$$

The reason for this seemingly mysterious choice of u and v is that $\sigma u = \omega u$ and $\sigma v = \omega^2 v$ and so, since σ can be thought of as an automorphism, $\sigma u^3 = (\sigma u)^3 = \omega^3 u^3 = u^3$ and similarly $\sigma v^3 = (\sigma^2 v)^3 = \omega^6 v^3 = v^3$. Hence σ fixes both u^3 and v^3 (the factors of 1/3 are merely for later convenience). Since σ generates A_3, all elements of A_3 must fix both u^3 and v^3. Furthermore, since K is a splitting field, K/L is a Galois extension and hence the fixed field of $\mathrm{Gal}(K/L)$ is precisely L. Thus we know that u^3 and v^3 are both in L.

A straightforward, albeit lengthy, calculation reveals that

$$\begin{aligned} 27u^3 &= x_1^3 + x_2^3 + x_3^3 - \tfrac{3}{2}(x_1^2x_2 + x_1x_2^2 + x_1^2x_3 + x_1x_3^2 + x_2^2x_3 + x_2x_3^2) \\ &\quad + 6x_1x_2x_3 + \tfrac{3}{2}(2\omega + 1)(x_1 - x_2)(x_1 - x_3)(x_2 - x_3), \end{aligned}$$

with the equivalent formula for v^3 obtained by swapping x_2 and x_3, which merely changes the sign of the final term. Hence, since $2\omega = -1 + \mathrm{i}\sqrt{3}$,

$$u^3 = -\tfrac{1}{2}q + \tfrac{1}{18}\mathrm{i}\sqrt{3}\,\delta \quad \text{and} \quad v^3 = -\tfrac{1}{2}q - \tfrac{1}{18}\mathrm{i}\sqrt{3}\,\delta, \tag{6.9}$$

which are both clearly in $L = F(\delta)$. We also have

$$9uv = x_1^2 + x_2^2 + x_3^2 - (x_1x_2 + x_1x_3 + x_2x_3) = -3p,$$

which is non-zero by assumption. This shows that there is just one choice of v once we have chosen one of the three cube roots for u (or vice versa). It follows from (6.8) that

$$x_1 = u + v, \quad x_2 = \omega u + \omega^2 v, \quad x_3 = \omega^2 u + \omega v,$$

where we have made use of $1 + \omega + \omega^2 = 0$. Substituting from (6.9) leads to Cardano's formula.

Another way to describe this, which mirrors an approach we will take to the quintic, is as follows. From (6.8), we can define ℓ_0, ℓ_1, ℓ_2 via

$$27u^3 = \ell_0 + \ell_1\omega + \ell_2\omega^2,$$
$$27v^3 = \ell_0 + \ell_2\omega + \ell_1\omega^2,$$

and if we substitute from (6.8), this yields

$$\ell_0 = x_1^3 + x_2^3 + x_3^3 + 6x_1x_2x_3 = -9q,$$
$$\ell_1 = 3(x_1x_2^2 + x_1^2x_3 + x_2x_3^2),$$
$$\ell_2 = 3(x_1^2x_2 + x_1x_3^2 + x_2^2x_3),$$

with $\ell_0 + \ell_1 + \ell_2 = (x_1 + x_2 + x_3)^3$, which is zero for the depressed cubic. With the aid of (6.7) we can show that the numbers ℓ_1 and ℓ_2 satisfy

$$\ell_1 + \ell_2 = 9q, \quad \ell_1\ell_2 = 9(9q^2 + p^3).$$

It follows that ℓ_1 and ℓ_2 are roots of a quadratic with coefficients in F. Both of ℓ_1 and ℓ_2 are fixed by $\sigma = (x_1x_2x_3)$ but we have

$$\tau\ell_1 = \ell_2, \quad \tau\ell_2 = \ell_1,$$

for any transposition τ and hence that ℓ_1 and ℓ_2 are in $F(\delta)$.

Solving the quadratic explicitly yields

$$\ell_{1,2} = \tfrac{1}{2}\,(9q \pm 3\delta)$$

and hence

$$27u^3 = -9q + \tfrac{9}{2}q(\omega + \omega^2) + \tfrac{3}{2}(\omega - \omega^2)\delta.$$

Changing δ to $-\delta$ gives the equivalent expression for v^3. Thus, with our specific choice of ω, we obtain (6.9) as before.

If the coefficients of the cubic are real then we can make some links between the Galois group and the number of real roots. When $f \in \mathbb{R}[x]$, we know (see Sect. 3.2) that f has either three real roots or one real root with the other two roots being a complex conjugate pair. In the latter case, there is an element of the Galois group which swaps the conjugate roots and leaves the real root fixed. This permutation is a transposition and hence odd, which is not possible if $G_f = A_3$. Hence, if $G_f = A_3$, then f has three real roots. However, we cannot infer anything about G_f from the fact that f has three real roots.

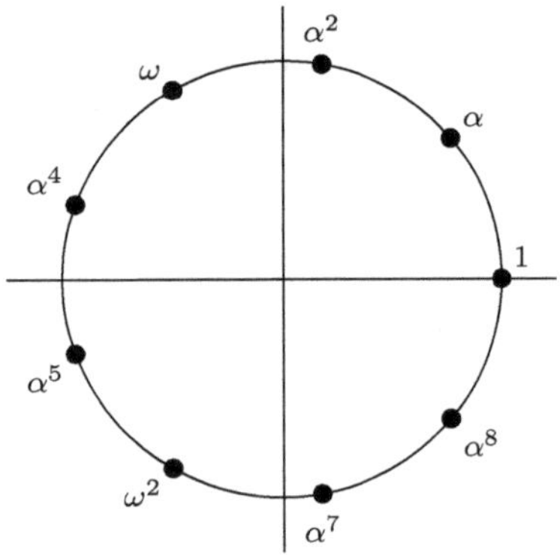

Fig. 6.1 The 9th roots of unity. Here $\alpha = e^{2\pi i/9}$ and $\omega = \alpha^3$. The diagram illustrates that $\alpha + \alpha^8$, $\alpha^2 + \alpha^7$ and $\alpha^4 + \alpha^5$ are all real, being the sum of a complex number and its complex conjugate

Example 6.16 In Example 6.14 we showed that the Galois group of $x^3 - 3x + 1 \in \mathbb{Q}[x]$ is A_3. Hence it must have three real roots. The solutions can be determined from the formulas above, with $p = -3$, $q = 1$, $\delta = 9$ and we find that $u^3 = \omega$ and $v^3 = \omega^2$, with $uv = 1$. If we let α be such that $\alpha^3 = \omega$, then we have $u = \alpha$, $v = \alpha^8$ and the three roots to the cubic are

$$\alpha + \alpha^8, \qquad \alpha^4 + \alpha^5, \qquad \alpha^7 + \alpha^2. \tag{6.10}$$

To see that these quantities are all real, note that α is a 9th root of unity, as well as a cube root of ω. Hence, following the example in Fig. 3.1, we can plot powers of α as equally spaced points around the unit circle in the complex plane. This is shown in Fig. 6.1 and it is apparent that each of the quantities in (6.10) is the sum of a number and its complex conjugate, and hence real. ♦

Finally, we turn to quartics. Let $f(x) = x^4 + bx^3 + cx^2 + dx + e \in F[x]$, with $F = \mathbb{Q}(b, c, d, e)$ and let $K = F(x_1, x_2, x_3, x_4)$ be its splitting field. As we did for cubics, we will collect together some useful results relating to symmetric polynomials in the roots that will make use of later on:

$$\begin{aligned} x_1 + x_2 + x_3 + x_4 &= -b, \\ x_1x_2 + x_1x_3 + x_1x_4 + x_2x_3 + x_2x_4 + x_3x_4 &= c, \\ x_1x_2x_3 + x_1x_2x_4 + x_1x_3x_4 + x_2x_3x_4 &= -d, \\ x_1x_2x_3x_4 &= e, \\ x_1^2 + x_2^2 + x_3^2 + x_4^2 &= b^2 - 2c, \end{aligned} \tag{6.11}$$

$$x_1^2x_2^2x_3^2 + x_1^2x_2^2x_4^2 + x_1^2x_3^2x_4^2 + x_2^2x_3^2x_4^2 = d^2 - 2ce,$$

$$x_1^2x_2^2 + x_1^2x_3^2 + x_1^2x_4^2 + x_2^2x_3^2 + x_2^2x_4^2 + x_3^2x_4^2 = c^2 - 2bd + 2e.$$

If f is reducible then either it factors into a linear factor and a cubic (reducible or otherwise), in which case the Galois group is just the Galois group of the cubic, or it factors into two irreducible quadratic factors. In the latter case the splitting field is $F(\delta_1, \delta_2)$, where δ_1^2 and δ_2^2 are the discriminants of the two quadratic factors. If $\delta_1 = a\delta_2, a \in F$, then the splitting field is simply the quadratic extension $F(\delta_1)$ and $G_f \cong C_2$. Otherwise we have the situation from Example 6.9, with $G_f \cong V_4$.

Assume now that f is irreducible over F. Certainly f has no roots in F, but that is not sufficient to guarantee irreducibility since f may be reducible and still have no roots in F; see Example 3.2. We know from property P2 that G_f must be a transitive subgroup of S_4 (with $|G_f|$ a multiple of 4). This leads to the following options (see Sect. 4.4):

$$S_4, \text{ order } 24; \; A_4, \text{ order } 12;$$

$$D_8, \text{ order 8 (three possibilities)};$$

$$C_4, \text{ order 4 (three possibilities)};$$

$$V_4, \text{ order } 4.$$

The only proper normal subgroups of S_4 are A_4 and V_4 and we also have $V_4 \lhd A_4$.

Our approach to understanding quartics will be to examine the intersection of G_f and V_4 (i.e. those elements which are in both groups, written $G_f \cap V_4$, and which is a subgroup of both of them) and we denote this group by H. We begin by establishing that H is a normal subgroup of G_f using the result that N is a normal subgroup of G if and only if $gNg^{-1} = N$ for any $g \in G$ (see Sect. 4.4).

Let $h \in H$ and $\sigma \in G_f$. Since h and σ are both elements of G_f, we know that $\sigma h \sigma^{-1}$ is in G_f by virtue of the fact that G_f is a group. Similarly, since $h \in V_4$ and $\sigma \in S_4$, we know that $\sigma h \sigma^{-1}$ is in V_4, because V_4 is normal in S_4. Hence $\sigma h \sigma^{-1} \in H$ and it follows that $H \trianglelefteq G_f$. Also, since H is a subgroup of V_4, we know that $|H| = 1, 2,$ or 4.

Let L be the fixed field of H. Since $H \trianglelefteq G_f$, the fundamental theorem of Galois theory tells us that L/F is a Galois extension with Galois group $\mathrm{Gal}(L/F) \cong G_f/H$. We also have $\mathrm{Gal}(K/L) \cong H$. The Galois correspondence is illustrated below:

$$\begin{array}{ccccc} F & \subseteq & L & \subseteq & K \\ \updownarrow & & \updownarrow & & \updownarrow \\ G_f & \trianglerighteq & H & \trianglerighteq & \{id\} \end{array}$$

We have already noted the importance in the theory of the quartic of the quantities

$$y_1 = x_1x_2 + x_3x_4, \quad y_2 = x_1x_3 + x_2x_4, \quad y_3 = x_1x_4 + x_2x_3. \tag{6.12}$$

The subgroup of S_4 that fixes y_i is $D_8^{(i)}$ $(i = 1, 2, 3)$ and the intersection of these three subgroups is V_4, i.e.

$$D_8^{(1)} \cap D_8^{(2)} \cap D_8^{(3)} = V_4.$$

Hence the subgroup of G_f that fixes all three of y_1, y_2 and y_3 is $H = G_f \cap V_4$. There are a few formalities to take care of, but it follows that the fixed field of H is

$$L = F(y_1, y_2, y_3)$$

and hence the fixed field of H is the splitting field for a cubic polynomial with roots y_1, y_2, y_3. The Galois group for that polynomial is then $\mathrm{Gal}(L/F) \cong G_f/H$.

Since elements of S_4 (acting on $\{x_1, x_2, x_3, x_4\}$) permute the numbers y_1, y_2, y_3 given in (6.12), we know that the elementary symmetric polynomials

$$y_1 + y_2 + y_3, \quad y_1y_2 + y_1y_3 + y_2y_3, \quad y_1y_2y_3$$

are fixed by S_4; they are symmetric polynomials in x_1, x_2, x_3, x_4 and hence in F. Lengthy calculations (using (6.11)) reveal that

$$y_1 + y_2 + y_3 = c, \quad y_1y_2 + y_1y_3 + y_2y_3 = bd - 4e, \quad y_1y_2y_3 = d^2 + b^2e - 4ce$$

and thus that y_1, y_2, y_3 are the roots of

$$g(y) = y^3 - cy^2 + (bd - 4e)y + 4ce - b^2e - d^2,$$

which is the cubic (2.13) from Ferrari's method.

Alternatively, we could have chosen the quantities

$$\tilde{y}_1 = (x_1 + x_2)(x_3 + x_4), \quad \tilde{y}_2 = (x_1 + x_3)(x_2 + x_4), \quad \tilde{y}_3 = (x_1 + x_4)(x_2 + x_3),$$

which again are invariant under the action of V_4. In fact

$$\tilde{y}_1 = y_2 + y_3, \quad \tilde{y}_2 = y_1 + y_3, \quad \tilde{y}_3 = y_1 + y_2.$$

The elementary symmetric polynomials in $\tilde{y}_1, \tilde{y}_2, \tilde{y}_3$ must again lie in F and we find that the cubic equation with roots $\tilde{y}_1, \tilde{y}_2, \tilde{y}_3$ is

$$\tilde{g}(y) = y^3 - 2cy^2 + (c^2 + bd - 4e)y + d^2 + b^2e - bcd.$$

Note that for the depressed quartic, where $b = 0$, this reduces to

$$y^3 - 2cy^2 + (c^2 - 4e)y + d^2.$$

If we identify y with $-s^2$, this is the same (up to a factor of -1) as the auxiliary cubic (2.18) that arises in Descartes' method. Clearly, $F(\tilde{y}_1, \tilde{y}_2, \tilde{y}_3) = F(y_1, y_2, y_3)$ and so the Galois groups of g and $\tilde{g}$ are the same. We also noted in Sect. 4.3 that the discriminants of g and $\tilde{g}$ are the same and identical to the discriminant $\Delta = \delta^2$ of the quartic from which they are derived.

Collecting information on the orders of the various groups, we note that

$$\begin{gathered} |G_f| = |G_f/H|\,|H| = |G_g|\,|H|, \\ |H| = 1, 2, \text{ or } 4, \text{ and } |G_f| \text{ is multiple of } 4. \end{gathered} \tag{6.13}$$

There are a number of different cases to consider, the first of which is whether or not the cubic g is irreducible over F (which is equivalent to saying that none of y_1, y_2, y_3 are in F).

We know from (6.6) that if g is irreducible over F, then $G_g \cong G_f/H$ is isomorphic to either S_3 or A_3 depending on whether $\delta \notin F$ or $\delta \in F$, respectively. Similarly, from Proposition 6.4 (P3), G_f is a subgroup of A_4 if and only if $\delta \in F$. It then follows from (6.13) that G_f is either S_4 or A_4 depending on whether $\delta \notin F$ or $\delta \in F$, respectively. Conversely, if $G_f = S_4$ or A_4 then (6.13) shows that $|G_g|$ must be a multiple of 3. This in turn implies that g must be irreducible. Hence we have shown that

- $G_f = S_4$ if and only if g is irreducible over F and $\delta \notin F$,
- $G_f = A_4$ if and only if g is irreducible over F and $\delta \in F$.

In either of these cases, the solution of the quartic will involve the extraction of cube roots.

On the other hand, if g is reducible over F, then we will be able to express the roots of the quartic using only square roots. Either g splits completely into three linear factors or it splits into a linear and an irreducible quadratic factor. In the first case we have $F(y_1, y_2, y_3) = F$ and hence $G_f = H$ (look back at the Galois correspondence) and so $G_f = V_4$. The same argument works in reverse and so we have that

- $G_f = V_4$ if and only if g splits completely over F.

It is usually more convenient to express this as a condition on δ. If g splits completely over F, then clearly $\delta \in F$. Conversely, if $\delta \in F$ than since f is irreducible we know that $G_f \leq A_4$, which given that $|G_f|$ is a multiple of 4 gives two possibilities, A_4 or V_4. However, we showed above that $G_f = A_4$ implies that

g is irreducible, so we must have $G_f = V_4$ and hence that g splits completely over F. We have thus shown that

- $G_f = V_4$ if and only if g is reducible over F and $\delta \in F$.

Finally, we need to consider the case where g splits into a linear and an irreducible quadratic factor over F. In this case $|G_g| = 2$ and so, from (6.13), we see that $|G_f| = 4$ or 8. G_f cannot be V_4 as this only happens when g splits completely over F, so either $G_f \cong D_8$ or $G_f \cong C_4$. One way to determine which it is, is to construct two quadratic polynomials q_1 and q_2, both in $F[x]$, the first having roots x_1x_2 and x_3x_4 and the second $x_1 + x_2$ and $x_3 + x_4$. These quadratics are, respectively,

$$q_1 = x^2 - y_1x + e \quad \text{and} \quad q_2 = x^2 + bx + c - y_1.$$

Denote the discriminants of q_1 and q_2 by $\Delta_1 = \delta_1^2$ and $\Delta_2 = \delta_2^2$:

$$\Delta_1 = y_1^2 - 4e, \quad \Delta_2 = b^2 - 4(c - y_1).$$

We will omit the details, but it can be shown that

- $G_f \cong C_4$ if and only if g has a single root y_1 in F and q_1 and q_2 both split over L, i.e. both $\delta_1\delta$ and $\delta_2\delta$ are in F,
- $G_f \cong D_8$ if and only if g has a single root y_1 in F, but q_1 and q_2 do not both split over L, i.e. at least one of $\delta_1\delta$ and $\delta_2\delta$ is not in F.

We conclude this section with some examples of the determination of the Galois groups of quartics.

Example 6.17 Biquadratic case. Let $f(x) = x^4 + cx^2 + e \in F[x]$ be irreducible. Then

$$g(y) = y^3 - cy^2 - 4ey + 4ce = (y - c)(y^2 - 4e).$$

Furthermore, $\Delta = 16e(c^2 - 4e)^2$. Hence,

$$G_f = V_4 \quad \text{if } \sqrt{e} \in F.$$

Otherwise, we construct $q_1(x) = x^2 - cx + e$ and $q_2(x) = x^2$ for which $\Delta_1 = c^2 - 4e$ and $\Delta_2 = 0$. Clearly $\delta_2\delta = 0 \in F$, so the nature of the Galois

(continued)

Example 6.17 (continued)

group hinges on whether $\delta_1\delta = \sqrt{16e(c^2-4e)^3} = 4(c^2-4e)\sqrt{e(c^2-4e)}$ is in F or not. Thus

$$G_f \cong \begin{cases} C_4 & \text{if } \sqrt{e} \notin F \text{ and } \sqrt{e(c^2-4e)} \in F, \\ D_8 & \text{if } \sqrt{e} \notin F \text{ and } \sqrt{e(c^2-4e)} \notin F. \end{cases}$$

Note that Example 6.10 has $c = -4$ and $e = 2$ so we get $e(c^2 - 4e) = 4^2$ and $G_f \cong C_4$. Also, if $c = 0$ and $\sqrt{e} \notin F$ as in Example 6.11, we have $G_f \cong D_8$ since $\sqrt{-4e^2} \notin F$. ♦

Example 6.18 Palindromic case. Let $f(x) = x^4+bx^3+cx^2+bx+1 \in F[x]$ be irreducible. First we note that the symmetry in the coefficients implies that

$$x^4 f(1/x) = f(x)$$

and hence if α is a solution, so is $1/\alpha$. We can therefore take the roots as

$$x_1 = \alpha, \quad x_2 = 1/\alpha \quad x_3 = \beta \quad x_4 = 1/\beta.$$

The associated cubic is

$$g(y) = y^3 - cy^2 + (b^2-4)y - 2b^2 + 4c$$

and since we know that $x_1x_2 + x_3x_4$ is a root of g we deduce that 2 is a root. Factorising gives

$$g(y) = (y-2)(y^2 + (2-c)y + b^2 - 2c)$$

and we can calculate $\Delta = (8 + b^2 - 4c)^2 D$, where

$$D = (2 - 2b + c)(2 + 2b + c) = (c+2)^2 - 4b^2.$$

Hence

$$G_f = V_4 \quad \text{if } \sqrt{D} \in F.$$

(continued)

Example 6.18 (continued)

On the other hand, if $\sqrt{D} \notin F$, we construct $q_1(x) = x^2 - 2x + 1 = (x-1)^2$, with zero discriminant, and $q_2(x) = x^2 + bx + c - 2$, for which $\Delta_2 = 8 + b^2 - 4c$. Hence

$$G_f \cong \begin{cases} C_4 & \text{if } \sqrt{D} \notin F \text{ and } \sqrt{D(8+b^2-4c)} \in F, \\ D_8 & \text{if } \sqrt{D} \notin F \text{ and } \sqrt{D(8+b^2-4c)} \notin F. \end{cases}$$ ♦

Example 6.19 Let $f(x) = x^4 + px + p \in \mathbb{Q}[x]$, where p is prime, which is irreducible by Eisenstein's criterion and has discriminant $\Delta = p^3(256 - 27p)$. The associated cubic is

$$g(y) = y^3 - 4py - p^2$$

and the rational root test tells us that any roots of g in $\mathbb{Q}$ must be integers which divide $-p^2$, i.e. $\pm 1,\ \pm p,\ \pm p^2$. We have

$$\begin{aligned} g(1) &= 1 - 4p - p^2, & g(-1) &= -1 + 4p - p^2, \\ g(p) &= p^2(p-5), & g(-p) &= p^2(3-p), \\ g(p^2) &= p^2(p^4 - 4p - 1), & g(-p^2) &= -p^2(p^4 - 4p + 1) \end{aligned}$$

and the only primes for which any of these vanish are $p = 3$ or $p = 5$. Thus unless $p = 3$ or $p = 5$ the cubic is irreducible. For $p \geq 11$, the discriminant is negative and for $p = 2$ and $p = 7$ it is not the square of an integer. Hence, for $p \neq 3, 5$ we have $G_f = S_4$.

For $p = 3$, we note that -3 is a root of g and

$$y^3 - 12y - 9 = (y+3)(y^2 - 3y - 3)$$

with splitting field $\mathbb{Q}(\sqrt{21})$. We have $q_1 = x^2 + 3x + 3$ and $q_2 = x^2 + 3$, neither of which has any real roots and so is irreducible over $\mathbb{Q}(\sqrt{21})$. Thus

$$\mathrm{Gal}(x^4 + 3x + 3, \mathbb{Q}) \cong D_8.$$

For $p = 5$, we have that 5 is a root of g and

$$y^3 - 20y - 25 = (y-5)(y^2 + 5y + 5),$$

(continued)

Example 6.19 (continued)

with splitting field $\mathbb{Q}(\sqrt{5})$. We have $q_1 = x^2 - 5x + 5$ and $q_2 = x^2 - 5$, both of which split over $\mathbb{Q}(\sqrt{5})$. Thus

$$\mathrm{Gal}(x^4 + 5x + 5, \mathbb{Q}) \cong C_4.$$

♦

There Is No General Quintic Formula

7

So far, we have given a fairly detailed account of the roots of polynomials, described the basic building blocks of Galois theory and shown how these ideas apply to quadratics, cubics and quartics. For quadratics, there is a simple formula which involves taking a single square root and once we admit complex numbers into our lexicon this always yields two solutions, one for each of the possible square roots. Cubic equations have a much richer structure; in general we need to take a cube root of an expression which contains a square root. This is related, in a way that we will make precise shortly, to the fact that the Galois group of a general cubic is S_3 which has order 6 and we can reduce 6 to 1 by first dividing by 2 (corresponding to a square root) and then by 3 (a cube root). We also found that the complex cube root of unity, ω, plays a crucial role, even when the polynomial and all its roots are real. For quartics things are more complicated, but no fundamentally new concepts come into play. The nature of the solution is a consequence of the factorisation $|S_4| = 24 = 2 \times 2 \times 2 \times 3$ and in general involves three layers of square roots and a cube root.

We now wish to turn our attention to finding the roots of quintic polynomials. Naively we might suspect that since $|S_5| = 5 \times |S_4|$, we will simply have to add fifth roots into the mix and everything will work nicely. But this is not the case and Galois theory will enable us to see why. However, the fact that there is no formula for the general quintic involving only a finite number of the operations $+, -, \times, \div$ and the extraction of roots (known as a **solution in radicals**) was known before Galois had any inkling of the group structure of the roots of polynomials.

The first person to give a clear statement that the quintic is not solvable in radicals was the Italian physician and mathematician Paolo Ruffini, who published five versions of this result between 1799 and 1813. His work, which built heavily on the foundations laid by Lagrange, was commented on by his contemporaries but perceived to be lacking in clarity, and it had little impact. Ruffini's proofs were incomplete though; he made a critical assumption without the necessary justification and it was left to the Norwegian mathematician Abel to provide the first complete

C. M. Linton, *Quintic Equations and How to Solve Them*,
https://doi.org/10.1007/978-3-032-01658-4_7

proof in 1824, with a significantly expanded exposition appearing in the first volume of a journal founded by August Leopold Crelle in 1826. Ruffini's work is described in detail in [3], an article which contains a great deal of insightful historical detail concerning attempts to understand the quintic. For details of Abel's proof, see, for example, [33], [59], [11, Lesson 10.3], and [31, Chapter 9]. Crelle's journal has a distinguished history and is still published today. Abel, who like Galois died very young (at 26) appears to have been unaware of Ruffini's contributions at the time he was developing his ideas, though he later wrote [33]:

> The first and, if I am not mistaken, the only one who, before me, has sought to prove the impossibility of the algebraic solution of general equations is the mathematician Ruffini. But his memoir is so complicated that it is very difficult to determine the validity of his argument. It seems to me that his argument is not completely satisfying.

On the other hand, we know from Galois' manuscripts that he was aware of the work of both Ruffini and Abel.

Clearly then, the full machinery of Galois theory is not necessary to establish this remarkable result. Moreover, it wasn't Galois' ideas that encouraged people like Ruffini and Abel to try and prove that the general quintic cannot be solved by radicals. So why did they suspect that the search for a quintic formula was futile?

The eighteenth century saw a gradual shift in attitudes amongst mathematicians towards the solution of equations with degree higher than four. In the seventeenth and early eighteenth centuries, when those such as James Gregory, Gottfried Leibniz and Ehrenfried Walther von Tschirnhaus were trying to find the way to crack the quintic, they were doing so in the full belief that a general formula, akin to that for the cubic and quartic, was out there waiting to be discovered. By mid-century, the language being used was somewhat less categorical and towards the end of the eighteenth century, Gauss was openly doubting that any such formula existed. At the time when Ruffini was attempting to prove the impossibility of solving quintics using radicals, the prevailing view was probably that the calculations involved in understanding the nature of the roots of quintic polynomials were so computationally onerous that even if a solution was out there, it may well be out of reach. One thing Gauss did achieve was to give the first complete proof that the nth roots of unity can always be expressed using a finite number of the basic operations of arithmetic and root extractions. In other words, the equation $x^n - 1 = 0$, for any positive integer n, is always solvable in radicals; see Sect. 3.3.

Throughout the eighteenth century, a general quintic formula was a highly sought after prize that tempted many great minds to devote considerable effort to the search. Even though the goal remained elusive, all sorts of fruitful ideas were generated in the process. Notable contributions were made by Abraham De Moivre, Leonhard Euler, Étienne Bézout, Gian Franceso Malfatti, Alexandre Vandermonde and Edward Waring. But it was Lagrange, working in the 1760s and 1770s, who collected the ideas of his predecessors, improved on them, and began to create a systematic theory of equations.

7.1 Resolvents

Fundamental to Lagrange's approach is the notion of a **resolvent**, which is a polynomial in the roots of an equation chosen to have specific properties. If we let ψ be a polynomial in $x_1, \ldots, x_n$ with coefficients in F, then under the action of the symmetric group S_n, ψ will take a number r of different values. The maximum possible value of r is $n!$, one for each element of S_n, but it may be less. The set of values $\{\psi_1, \ldots, \psi_r\}$ is called the **orbit** of ψ under the action of S_n. Lagrange realised that the length of the orbit, r, must divide into $n!$ and this is in fact his version of the statement that we now refer to as Lagrange's theorem in group theory; see, for example, [33, pp. 53–54].

For a given ψ, we can construct the **resolvent polynomial**

$$\theta(x) = (x - \psi_1) \cdots (x - \psi_r),$$

the coefficients of which are symmetric polynomials in $\psi_1, \ldots, \psi_r$ and hence, since any element of S_n just reorders $\psi_1, \ldots, \psi_r$, symmetric polynomials in $x_1, \ldots, x_n$. This shows that $\theta \in F[x]$.

Example 7.1 Take $n = 4$ and set $\psi = x_1x_2+x_3x_4$. We have already seen (for example in Example 6.1) that ψ takes 3 values under the action of S_4. Hence θ is a cubic which can be solved using radicals. We showed in Sect. 6.5 that this is the cubic (2.13) from Ferrari's method. Hence this cubic is often referred to as a resolvent cubic for the associated quartic. ♦

Lagrange's approach to finding the roots of a polynomial f was to seek resolvents ψ with two key properties. First, the resolvent polynomials that they generate have lower degree than f, or at least are decomposable into other polynomials with suitably low degree; and second, that the values of the roots of f can be determined, preferably easily, from knowledge of the roots of the resolvent polynomial.

To this end, he introduced what is now known as the **Lagrange resolvent**. For an nth degree equation, writing $\mu \neq 1$ as an nth root of unity, we can construct

$$\psi(x_1, \ldots, x_n) = x_1 + \mu x_2 + \ldots \mu^{n-1} x_n.$$

If we apply the permutation $\sigma = (x_1x_2 \cdots x_n) \in S_n$ to ψ and then multiply by μ, we will recover ψ. We can repeat this process and so

$$\psi = \mu\sigma\psi = \mu^2\sigma^2\psi = \cdots = \mu^{n-1}\sigma^{n-1}\psi$$

and it follows that $\psi^n = \sigma\psi^n = \cdots = \sigma^{n-1}\psi^n$. Thus the n distinct powers of σ fix ψ^n and it can be shown that any element of S_n that fixes ψ^n must be a power of σ. This leads to the conclusion that the number of values taken by ψ^n under the action of S_n, and thus the degree of the resolvent polynomial for ψ^n, is $n!/n = (n-1)!$. Moreover, if this equation can be solved so that ψ is known in terms of the coefficients of the original polynomial, then it is relatively straightforward to recover $x_1, \ldots, x_n$. For example, we find that

$$nx_1 = x_1 + \cdots + x_n + \psi(x_1, x_2, x_3 \ldots, x_{n-1}, x_n)$$
$$+ \psi(x_1, x_3, \ldots, x_{n-1}, x_n, x_2) + \cdots + \psi(x_1, x_n, x_2, x_3, \ldots, x_{n-1}),$$

where we have cycled the roots $x_2, \ldots, x_n$ in the $n-1$ expressions for ψ. When summing the terms on the right-hand side, each of the roots, apart from x_1, is multiplied by the factor $1 + \mu + \cdots + \mu^{n-1}$, which is zero.

Example 7.2 For $n = 3$, this is exactly as we did in Sect. 6.5. In terms of the notation that we have been using for cubics, see (6.8), we recognise that the Lagrange resolvent ψ is our $3v = x_1 + \omega x_2 + \omega^2 x_3$ and ψ^3 takes $(3-1)! = 2$ values under the action of S_3 (namely $27v^3$ and $27u^3$, where $3u = x_1 + \omega^2 x_2 + \omega x_3$). We showed how to express u^3 and v^3 in terms of the coefficients of the original cubic (this involves solving a quadratic equation) and then the roots are $x_1 = u + v$, $x_2 = \omega u + \omega^2 v$ and $x_3 = \omega^2 u + \omega v$.

We can also consider the resolvent equation for ψ rather than ψ^3, though this may not at first seem to fit the bill since ψ takes 6 values under the action of S_3 and thus we will end up with a sextic equation. However, if we compute this sextic we find that it is in fact a quadratic in x^3 and so leads to exactly the same solutions. ♦

Example 7.3 For $x^4 + bx^3 + cx^2 + dx + e$, we can take the Lagrange resolvent $\psi = x_1 - x_2 + x_3 - x_4$, in which we have set $\mu = -1$ (one of the fourth roots of unity). We can show that ψ takes 6 values under the action of S_4, namely $\pm\psi_1, \pm\psi_2, \pm\psi_3$, where

$$\psi_1 = x_1 - x_2 + x_3 - x_4, \quad \psi_2 = x_1 + x_2 - x_3 - x_4,$$
$$\psi_3 = x_1 - x_2 - x_3 + x_4.$$

The resolvent polynomial for ψ is the sextic

$$\theta(x) = (x^2 - \psi_1^2)(x^2 - \psi_2^2)(x^2 - \psi_3^2)$$

(continued)

Example 7.3 (continued)

and we can compute

$$\begin{aligned}\psi_1^2 &= x_1^2 + x_2^2 + x_3^2 + x_4^2 \\ &\quad + 2(-x_1x_2 + x_1x_3 - x_1x_4 - x_2x_3 + x_2x_4 - x_3x_4) \\ &= b^2 - 2c + 2(-y_1 + y_2 - y_3) \\ &= b^2 - 4c + 4y_2.\end{aligned}$$

Here $y_1 = x_1x_2 + x_3x_4$, $y_2 = x_1x_3 + x_2x_4$, $y_3 = x_1x_4 + x_2x_3$ as in Sect. 6.5 and we have used the fact that $y_1 + y_2 + y_3 = c$. Similarly

$$\psi_2^2 = b^2 - 4c + 4y_1 \quad \text{and} \quad \psi_3^2 = b^2 - 4c + 4y_3.$$

We can't choose the signs of ψ_1, ψ_2 and ψ_3 independently, since $\psi_1\psi_2\psi_3$ is a symmetric polynomial in $x_1, \ldots, x_n$ and hence has a given value in terms of the coefficients of the original quartic. In fact we can show that

$$\psi_1\psi_2\psi_3 = -b^3 + 4bc - 8d.$$

We know that y_1, y_2, y_3 are the solutions to the cubic (2.13) and then, since $x_1 + x_2 + x_3 + x_4 = -b$,

$$\begin{aligned}4x_1 &= -b + \psi_1 + \psi_2 + \psi_3, \\ 4x_2 &= -b - \psi_1 + \psi_2 - \psi_3, \\ 4x_3 &= -b + \psi_1 - \psi_2 - \psi_3, \\ 4x_4 &= -b - \psi_1 - \psi_2 + \psi_3.\end{aligned}$$

Even though the resolvent polynomial is of higher degree than 4, it has a structure that enables us to solve it. ♦

These examples do not give us information about the solutions to the cubic and quartic that we have not already unearthed, but it is the systematic nature of the approach that suggests ways to tackle higher-degree equations like the quintic. Through the study of equations of low degree, Lagrange hoped to find patterns and structures that could be generalised to higher degrees. In fact, in the case where n is prime, Lagrange showed how to use his resolvents to construct resolvent polynomials whose degree was $(n-2)!$. For the quintic this produces a resolvent of degree 6, but his attempts to simplify it led nowhere. Lagrange concluded that if a general formula exists for equations of degree higher than 4, then it must depend on some other connections between the roots. In 1770, he wrote [31]:

> It may seem reasonable to apply these principles to equations of the fifth and higher degrees, of which the solution is at present unknown, but such application requires too large a number of examinations and combinations whose success is very much in doubt. Thus, for the time being, we must give up this work. We hope to be able to return to it at another time. At this point we shall be satisfied with having set forth the foundations of a theory which appears to us to be new and general.

It was against this background that Ruffini decided to take a different tack. How might one prove that no general formula for solving quintic equations exists? This is a question of a fundamentally different character to that of trying to find such a formula. Finding something, however well hidden, is conceptually clear cut; once you have identified it you are done. But how might one establish the non-existence of an expression for the roots of a quintic polynomial involving only algebraic operations. It is clearly not possible to exhaust all possibilities as there are infinitely many. Rather than looking for a needle in a haystack, we want to show that it is not possible for there to be a needle in the haystack; how can we do that without combing through every possible mound of hay? In tackling a problem framed in terms of existence rather than, as was the norm for the time, in the form of a computation, Ruffini was truly breaking new ground and it is perhaps not that surprising that he was frequently accused of vagueness. After all, the language for expressing these concepts clearly and unambiguously was being developed at the same time as the ideas themselves.

A clue that the $n = 5$ case is qualitatively different from cubics and quartics can be seen in the fact that:

Proposition 7.1 *It is not possible to find a resolvent $\psi(x_1, \ldots, x_5)$ that takes 3 or 4 values under the action of S_5.*

It is always possible to find a resolvent in $x_1, \ldots, x_n$ which takes a single value under the action of S_n, any symmetric polynomial will do. For a separable polynomial, finding a resolvent that takes two values is also straightforward, since we can take the square root of the discriminant: $\delta = (x_1 - x_2)(x_1 - x_3) \cdots (x_{n-1} - x_n)$, which under the action of an element of S_n becomes one of $\pm\,\delta$. Also, the length of the orbit, r, of the resolvent $\psi = x_1$ is clearly n provided the roots are distinct. So for separable polynomials $r = 1, 2$ and n are always possible. For quartics, since there are three ways to split four objects into two pairs, we have $r = 3$ for $\psi = x_1x_2 + x_3x_4$ and this was crucial in reducing the general quartic to a cubic. But Ruffini was able to show through a painstaking set of calculations that for quintics, $r = 3$ and $r = 4$ are impossible. In the terminology of group theory, what Ruffini had discovered was that S_5 has no subgroups of index 3 or 4, i.e. order $120/3 = 40$ or $120/4 = 30$, something which is far from obvious.

While there are quite a few technicalities to overcome, and it was Abel who first managed to put everything together into what would now be considered an acceptable proof, this result of Ruffini's (generalised and brought to a much wider audience by Cauchy in 1812) is the key to proving that no general formula exists. Abel argued that the form of the solution that a general formula would have must be in the form of nested roots of expressions which are polynomials in the roots $x_1, \ldots, x_5$ and because of the absence of resolvents with $r = 3$ or 4, these roots must be square roots or fifth roots. The innermost root being a fifth root could be ruled out by comparing the number of possible values of certain expressions so this meant that it must be a square root. Then the next root up in the sequence of nested roots must be of degree 2 or 5, just as before. It can't be of degree 2 because then we would have a square root of a square root, which is a fourth root and therefore not allowed. But this root being a fifth root could be ruled out in the same way that the impossibility of the innermost root being a fifth root was. This leads to the conclusion that there is no possible expression for the roots of a general quintic involving just (finitely many) algebraic operations. The lack of a general formula for a quintic immediately implies the lack of such a formula for any higher order polynomial since a quintic is a special case of any polynomial of degree higher than 5 and the same formula would apply to the quintic after setting some of the coefficients to zero.

While not everyone was immediately willing to accept this counterintuitive result, the impossibility of solving equations of degree higher than 4 by radicals was fairly quickly acknowledged following the publication of Abel's proof. Weaknesses in Abel's arguments were identified and new, clearer proofs were produced. But the core ideas that Abel had put into print were sound and the search for a general quintic formula was at an end; there isn't one.

As is often the case in mathematics, the closing of one door simply opens another. There is no formula that will solve a general quintic in the same way that there is for quadratics, cubics and quartics. But some quintics *can* be solved in terms of their coefficients, most obviously $x^5 - a$. So we are now faced with new questions. How do we determine whether a given quintic equation is solvable by radicals and, if it is, how might we go about solving it? The story certainly does not end with Abel's impossibility proof. This is where the power of Galois' approach comes into its own. First we will look at how the impossibility of solving a general quintic by radicals can be established using the framework of Galois theory and we will begin by making more precise what we actually mean by a solution in radicals.

7.2 Solutions in Radicals

Solutions in radicals are solutions which are obtained from the coefficients in the original equation using only a finite number of the operations $+$, $-$, $\times$, $\div$ and the extraction of roots. The quadratic formula, which ascribes the roots

$$\tfrac{1}{2}(-b \pm \delta), \quad \delta = \sqrt{b^2 - 4c}$$

to the polynomial $x^2 + bx + c$ is an example of a solution in radicals. With the base field $F = \mathbb{Q}(b, c)$ we see that the roots are contained in the field $F(\delta)$, with $\delta^2 \in F$. The field $F(\delta)$ is an example of what is known as a radical extension.

A field extension $F(\alpha)$ is a **radical extension** if $\alpha^m \in F$ for some positive integer m. We also say that $F(\alpha_1, \ldots, \alpha_n) = F(\alpha_1)(\alpha_2)\ldots(\alpha_n)$ is a radical extension if each of the successive extensions in the chain is a radical extension. The radicals themselves can always be chosen with m prime, since, for example, if we want to adjoin $\sqrt[6]{\alpha}$ we can first adjoin $\beta = \sqrt{\alpha}$ and then $\sqrt[3]{\beta}$.

Example 7.4 $\mathbb{Q}(\sqrt{2})$ is a radical extension since $(\sqrt{2})^2 \in \mathbb{Q}$. ♦

Example 7.5 $\mathbb{Q}\left(\sqrt{2}, \sqrt[3]{1+\sqrt{2}}\right)$ is a radical extension since $\left(\sqrt[3]{1+\sqrt{2}}\right)^3 \in \mathbb{Q}(\sqrt{2})$. ♦

However, a radical extension is not necessarily a Galois extension—we have already seen (in Example 6.6) that $\mathbb{Q}(\sqrt[3]{2})$ is not Galois. To rectify this, in any solution in radicals, before we adjoin a pth root for some prime p, we can adjoin a pth root of unity (other than 1) if it is not already present. (Adjoining a pth root is always a radical extension since it satisfies $x^p = 1$ and 1 is in all fields.) In this way, whenever we adjoin a root of $x^p - \alpha$, we adjoin all the roots of the equation. This will ensure that the extensions are Galois extensions.

Example 7.6 Cardano's formula for obtaining the three roots of $x^3 + px + q \in F[x]$ involves the radical extension $F(\sqrt{D}, \omega, u)$, where

$$D = \tfrac{1}{27}p^3 + \tfrac{1}{4}q^2 \quad \text{and} \quad u = \sqrt[3]{-\tfrac{1}{2}q + \sqrt{D}}.$$

Note that we do not need to adjoin $v = \sqrt[3]{-\frac{1}{2}q - \sqrt{D}}$ as well as u, because $uv = -p/3$. ♦

A **solution in radicals** to the equation $x^n + a_{n-1}x^{n-1} + \cdots + a_1x + a_0 = 0$ can be thought of as a process which starts with the base field $F = \mathbb{Q}(a_0, \ldots, a_{n-1})$ and proceeds by adjoining radicals until a field containing the roots $x_1, \ldots, x_n$ is obtained. To make this precise, we say that a polynomial $f \in F[x]$ with splitting field K is **solvable by radicals** if and only if there exists a tower of radical field extensions

$$F = F_0 \subseteq F_1 \subseteq \ldots \subseteq F_k = F(\alpha_1, \ldots, \alpha_k) = E, \tag{7.1}$$

where for each $j = 1, \ldots, k$, we have the radical extension $F_j = F_{j-1}(\alpha_j)$, with $\alpha_j^{p_j} \in F_{j-1}$ for some prime p_j, and such that $K \subseteq E$. It is an important feature that we require the splitting field K to be contained in the radical extension E rather than that E is itself the splitting field for f. As we have seen, for example, the solution in radicals to cubic polynomials $f \in \mathbb{Q}[x]$ in which all the roots are real, involves complex numbers and yet $K \subset \mathbb{R}$. The question of whether an irreducible polynomial can be solved using real radicals alone is answered in [13, Thm. 8.6.5]. A necessary, but not sufficient, condition is that the degree of the polynomial is a power of 2.

Let us assume that f is solvable by radicals and consider a general extension F_j/F_{j-1} from the tower (7.1). Let μ be a non-trivial p_jth root of unity. Then the roots of $x^{p_j} - \alpha_j^{p_j}$ are $\alpha_j, \alpha_j\mu, \alpha_j\mu^2, \ldots \alpha_j\mu^{p_j-1}$. It follows that $F_{j-1}(\alpha_j)$ is the splitting field for $x^{p_j} - \alpha_j^{p_j}$ over F_{j-1}. Hence each radical extension in the tower is a Galois extension. We also know that $\sigma \in \mathrm{Gal}(F_j/F_{j-1})$ must map α_j to one of the roots, i.e. $\sigma(\alpha_j) = \alpha_j\mu^q$ for some $q = 0, \ldots, p_j - 1$. The action of σ is thus equivalent to multiplication by a power of μ and it is then straightforward to show that if we have two elements of $\mathrm{Gal}(F_j/F_{j-1})$, σ and τ, then $\sigma\tau = \tau\sigma$ and hence $\mathrm{Gal}(F_j/F_{j-1})$ is abelian.

We thus have a tower of extensions, each one of which is a Galois extension and such that the Galois group of the extension is abelian. There is one final piece of the jigsaw that we need and this stems from the fact that it is not guaranteed that E/F is Galois, which is needed to be able to make use of the fundamental theorem of Galois theory in what follows. In order to get around this problem, we need to construct what is called the **Galois closure** of E over F, which is the defined as the smallest field E' which contains E and which is Galois over F. It is a somewhat technical exercise, but it can be shown that if E is a radical extension of F, then so is E' [13, Thm. 8.2.8].

We now appeal to the Galois correspondence to associate with our tower of fields a descending tower of groups

$$\mathrm{Gal}(E'/F) = G_0 \geq G_1 \geq \ldots \geq G_k = \{id\},$$

where $G_j = \mathrm{Gal}(E'/F_j)$. From the fundamental theorem of Galois theory we know that if $F_{j-1} \subseteq F_j \subseteq E'$, with E' a Galois extension of F_{j-1}, then F_j is a Galois extension of F_{j-1} if and only if G_j is a normal subgroup of G_{j-1} (see Sect. 6.4).

Hence in the case of a radical extension, where all the extensions are certainly Galois, we have

$$\mathrm{Gal}(E'/F) = G_0 \trianglerighteq G_1 \trianglerighteq \ldots \trianglerighteq G_k = \{id\}.$$

Thus Galois theory converts a statement about a tower of fields, each of which is a radical extension of its predecessor, into one about a chain of groups, each of which is a normal subgroup of the previous one.

7.3 Solvable Groups

Let us put our main task to one side temporarily and explore the structure of group chains like the one above, starting with the introduction of some helpful terminology. A **subnormal series** of a group G is a chain of subgroups each of which is a normal subgroup of the preceding one in the chain. In other words, we have

$$G = G_0 \triangleright G_1 \triangleright \ldots \triangleright G_m = \{id\}. \tag{7.2}$$

Because the subgroups in the chain are normal subgroups, we can form the quotient groups G_i/G_{i+1} corresponding to each link in the chain; these are called the **factors** of the series. It is not the case that the groups $G_2, \ldots, G_m$ are necessarily normal subgroups of G even though each group is a normal subgroup of the previous group in the series. If, in addition, it is true that $G \triangleright G_i$ for all the groups in the series, then the series is a **normal series**. Since every subgroup of an abelian group is normal (Proposition 4.2), it follows that if G is an abelian group, any subnormal series is also normal, but this is not the case in general. It may be possible to insert a group H into (7.2) and it remain a subnormal series. In other words

$$G = G_0 \triangleright \ldots G_i \triangleright H \triangleright G_{i+1} \ldots \triangleright G_m = \{id\}.$$

Such a series is called a **refinement** of (7.2).

A subnormal series is a **composition series** if each of the factors in the series is simple (meaning that they have no proper normal subgroups). This is the same as requiring that the series cannot be refined. Every group has a composition series; in fact a group may have more than one composition series, but the Jordan-Hölder theorem [13, §8.4] tells us that they are all equivalent in the sense that the set of factors in each series is the same up to isomorphism.

A group G is said to be **solvable** if it possesses a subnormal series such that each of the factors is abelian. If one of the factors is abelian but not simple the series can be refined and we can continue to do this until all the factors are simple and abelian. Since any subgroup of an abelian group is normal, a simple abelian group cannot have any proper subgroups. This implies that every simple abelian group is cyclic of prime order and hence a group G is solvable if and only if it possesses a composition series such that each of the factors is cyclic and of prime order. (See the remarks at the end of Sect. 4.4.)

Example 7.7 We have the composition series

$$S_3 \rhd A_3 \rhd \{id\},$$

with $|S_3| = 6$, $|A_3| = 3$. The factors are $S_3/A_3 \cong C_2$ and $A_3/\{id\} \cong C_3$ which are both simple, cyclic and of prime order. Hence S_3 is solvable. ♦

Example 7.8 We have the subnormal series

$$S_4 \rhd A_4 \rhd V_4 \rhd \{id\}, \tag{7.3}$$

with $|S_4| = 24$, $|A_4| = 12$, $|V_4| = 4$. The factors are $S_4/A_4 \cong C_2$, $A_4/V_4 \cong C_3$ and $V_4/\{id\} \cong V_4$, which are all abelian. Hence S_4 is solvable. However, this is not a composition series since V_4 is not simple. We can refine (7.3) to give the composition series

$$S_4 \rhd A_4 \rhd V_4 \rhd C_2 \rhd \{id\},$$

with factors isomorphic to C_2 (three times) and C_3. ♦

Three basic properties of solvable groups are [13, §8.1]:

Proposition 7.2

- *All abelian groups are solvable.*
- *If G is solvable then all subgroups of G are solvable.*
- *If G is solvable and $N \lhd G$, then G/N is solvable.*

It will come as no surprise that the reason for the term *solvable group* is directly related to the solvability by radicals of the associated polynomial. What we showed earlier was that a polynomial $f \in F[x]$ is solvable in radicals if and only if there is a field E' which is the Galois closure of a radical extension of F such that the splitting field K of f is contained in E'. If such a field E' exists, then the fundamental theorem of Galois theory leads to the conclusion that $\text{Gal}(E'/F)$ is a solvable group.

Now $F \subseteq K \subseteq E'$ by hypothesis, so we can let H be the subgroup of $\text{Gal}(E'/F)$ which fixes K. Since K is a splitting field, we know that K/F is a Galois extension and hence, from the fundamental theorem of Galois theory

$$G_f = \text{Gal}(K/F) \cong \text{Gal}(E'/F)/H.$$

It follows from the properties of solvable groups, that G_f is solvable. The converse can also be shown to be true. Thus, the polynomial $f \in F[x]$ is solvable by radicals if and only if G_f is a solvable group.

Example 7.9 The smallest simple non-abelian group is A_5, with $|A_5| = 60$ (the next smallest has order 168). Since A_5 is simple, we have the composition series

$$S_5 \triangleright A_5 \triangleright \{id\},$$

with $|S_5| = 120$. The factors are isomorphic to C_2 and A_5. Since A_5 is not abelian, S_5 is not solvable. Nor is A_5. All groups of order less than 60 are solvable. ♦

7.4 Implications for Quintic Equations

The Galois group of a general quintic, where there are no algebraic relations between the coefficients, is S_5. Since S_5 is not solvable we conclude that the general polynomial of degree 5 is not solvable by radicals. Of course, a specific polynomial of degree 5 may have a solution in radicals, e.g. $x^5 - a$.

The Galois group of any irreducible quintic must be a transitive subgroup of S_5 and these are easily listed. They are

- S_5, order 120. The whole group.
- A_5, the alternating group of order 60. The subgroup of S_5 consisting of all the even permutations.
- F_{20}, the Frobenius group of order 20. (This group goes by many different names; it is sometimes referred to as the metacyclic group of order 20 and often, as in [13], as the affine linear group $AGL(1, \mathbb{F}_5)$). There are 6 subgroups of S_5 that are isomorphic to F_{20}, all of which are of the form

$$\langle \sigma, \tau \rangle, \quad \sigma^5 = \tau^4 = id, \ \sigma\tau = \tau\sigma^2$$

so that the elements are $\sigma^i\tau^j$, $i = 0, 1, 2, 3, 4$, $j = 0, 1, 2, 3$. None of these groups is a subgroup of A_5. There are 24 possible 5-cycles in S_5 and each of the 6 conjugate subgroups isomorphic to F_{20} contains four of them. We will select a particular subgroup by taking $\sigma = (x_1x_2x_3x_4x_5)$ and then there are lots of choices for τ. We choose $\tau = (x_2x_3x_5x_4)$ as it will prove convenient that τ fixes x_1. Choosing one of the 6 possible instantiations of F_{20} has the effect of restricting the possible ordering of the roots $x_1, \ldots, x_5$. The order of the elements (the smallest power they need to be raised to in order to yield *id*) is given by

order	elements
1	*id*
2	$\tau^2, \sigma\tau^2, \sigma^2\tau^2, \sigma^3\tau^2, \sigma^4\tau^2$
4	$\tau, \sigma\tau, \sigma^2\tau, \sigma^3\tau, \sigma^4\tau, \tau^3, \sigma\tau^3, \sigma^2\tau^3, \sigma^3\tau^3, \sigma^4\tau^3$
5	$\sigma, \sigma^2, \sigma^3, \sigma^4$

No element of F_{20} (apart form *id*) acting on $x_1, \ldots, x_5$ fixes more than one of the roots (this is the property that characterises Frobenius groups).

- D_{10}, the dihedral group of order 10. The full symmetry group of a regular pentagon. There are 6 subgroups of S_5 that are isomorphic to D_{10}, all of which are subgroups of A_5. In fact, D_{10} is the intersection of A_5 and F_{20} and so our particular choice for F_{20} determines the choice of D_{10}.
- C_5, the cyclic group of order 5. There are 6 subgroups of S_5 that are isomorphic to C_5, all of which are subgroups of A_5. All are of the form $\{id, \sigma, \sigma^2, \sigma^3, \sigma^4\}$, $\sigma^5 = id$ and we will take $\sigma = (x_1x_2x_3x_4x_5)$ so that C_5 is a subgroup of F_{20} and of D_{10}.

Based on this list, we can immediately make some observations about the Galois group of a quintic $q \in F[x]$. Galois theory tells us that polynomial equations are solvable by radicals if and only if their Galois group is solvable and we have seen that S_5 and A_5 are not solvable but that any group with order less than 60 is solvable. We also know (Proposition 6.3) that if the discriminant Δ of a separable quintic is a square in F, the Galois group must be a subgroup of A_5, and conversely.

Since F_{20} is not contained within A_5, but D_{10} and C_5 are both subgroups of A_5 we see that if Δ is a not a square in F, then the Galois group is either S_5 (q not solvable) or F_{20} (q solvable). If Δ is a square in F then the Galois group is either A_5 (q not solvable) or one of D_{10} or C_5 (q solvable). This categorisation is illustrated in Table 7.1. Determining between D_{10} and C_5 is far from easy in general; we give some methods in Sect. 9.5.

Finally, we note that no element of A_5, F_{20}, D_{10} or C_5 is a transposition. If an irreducible quintic with real coefficients has three real roots, then the other two roots must be a complex conjugate pair and the automorphism $\phi(x + \mathrm{i}y) = x - \mathrm{i}y$ swaps these roots and so is a transposition. Hence such a quintic must have Galois group S_5 and is not solvable. This result also follows from a theorem of Galois (anticipated by Abel) that a necessary and sufficient condition for an irreducible polynomial of

Table 7.1 Galois groups of quintic polynomials $q \in F[x]$

G_q	Δ a square in F	Solvable by radicals
S_5	No	No
A_5	Yes	No
F_{20}	No	Yes
D_{10} or C_5	Yes	Yes

prime degree to be solvable is that once any two roots are known, the others can be expressed as rational functions of them [59] and [6, Thm. 9.4]. If two roots are real and the equation is solvable, then all the roots must be real. Thus we have:

Proposition 7.3 *If an irreducible quintic with real coefficients is solvable by radicals, then it has either one or five real roots.*

The presence of complex conjugate pairs of roots implies that complex conjugation is a non-trivial automorphism of the splitting field and this corresponds to the existence of an element of the Galois group of order 2. However, all elements of the cyclic group C_5 have order 5. Thus:

Proposition 7.4 *If a solvable, irreducible quintic with real coefficients has one real root, then its Galois group is either F_{20} or D_{10}.*

For the purposes of doing explicit calculations, we now list the elements of F_{20} explicitly. With $\sigma = (x_1x_2x_3x_4x_5)$ and $\tau = (x_2x_3x_5x_4)$, they are as follows. We have *id* and then five elements of order 2,

$$\tau^2 = (x_2x_5)(x_3x_4), \quad \sigma\tau^2 = (x_1x_5)(x_2x_4), \quad \sigma^2\tau^2 = (x_1x_4)(x_2x_3),$$
$$\sigma^3\tau^2 = (x_1x_3)(x_4x_5), \quad \sigma^4\tau^2 = (x_1x_2)(x_3x_5),$$

10 elements of order 4,

$$\tau = (x_2x_3x_5x_4), \quad \sigma\tau = (x_1x_3x_2x_5), \quad \sigma^2\tau = (x_1x_5x_3x_4),$$
$$\sigma^3\tau = (x_1x_2x_4x_3), \quad \sigma^4\tau = (x_1x_4x_5x_2),$$
$$\tau^3 = (x_2x_4x_5x_3), \quad \sigma\tau^3 = (x_1x_4x_3x_5), \quad \sigma^2\tau^3 = (x_1x_2x_5x_4),$$
$$\sigma^3\tau^3 = (x_1x_5x_2x_3), \quad \sigma^4\tau^3 = (x_1x_3x_4x_2),$$

and four elements of order 5,

$$\sigma = (x_1x_2x_3x_4x_5), \quad \sigma^2 = (x_1x_3x_5x_2x_4),$$
$$\sigma^3 = (x_1x_4x_2x_5x_3), \quad \sigma^4 = (x_1x_5x_4x_3x_2).$$

For reference, we note the following relations:

$$(\sigma\tau)^2 = \sigma^4\tau^2, \quad (\sigma^2\tau)^2 = \sigma^3\tau^2, \quad (\sigma^3\tau)^2 = \sigma^2\tau^2, \quad (\sigma^4\tau)^2 = \sigma\tau^2,$$
$$(\sigma\tau)^3 = \sigma^3\tau^3, \quad (\sigma^2\tau)^3 = \sigma\tau^3, \quad (\sigma^3\tau)^3 = \sigma^4\tau^3, \quad (\sigma^4\tau)^3 = \sigma^2\tau^3.$$

Example 7.10 We can now illustrate the Galois correspondence for the polynomial

$$f = x^5 - 2$$

as we did previously for $x^3 - 2$ and $x^4 - 2$ in Examples 6.12 and 6.13. We write $\alpha = \sqrt[5]{2}$ for convenience. The splitting field is $\mathbb{Q}(\alpha, \zeta)$ and the Galois group is

$$G_f = \mathrm{Gal}(\mathbb{Q}(\alpha, \zeta)).$$

The minimum polynomial of α has order 5 and that of ζ has order 4. Hence $|G_f| = 20$ and thus $G_f \cong F_{20}$. We can arrive at the same conclusion by noting that the discriminant is $\Delta = 50{,}000$ so that $\delta = 100\sqrt{5}$, which is not rational, and we therefore know that G_f is not a subgroup of A_5. Since f is solvable, this implies that $G_f \cong F_{20}$ as before. We label the roots in the obvious way:

$$x_1 = \alpha, \quad x_2 = \alpha\zeta, \quad x_3 = \alpha\zeta^2, \quad x_4 = \alpha\zeta^3, \quad x_5 = \alpha\zeta^4.$$

Any automorphism is uniquely determined by its effect on α and on ζ and we know that $\phi(\alpha)$ must be one of the 5 roots. Also we have that $\phi(\zeta)^5 = \phi(1) = 1$ and so $\phi(\zeta)$ is one of $\zeta, \zeta^2, \zeta^3, \zeta^4$. It cannot be 1 because $\phi(1) = 1$. Hence we can define automorphisms ϕ_j, $j = 1, \ldots, 20$, as follows:

$$\phi_j(\alpha) = \alpha\zeta^{j-1}, \quad j = 1, \ldots, 20,$$
$$\phi_j(\zeta) = \begin{cases} \zeta & j = 1, \ldots, 5, \\ \zeta^2 & j = 6, \ldots, 10, \\ \zeta^3 & j = 11, \ldots, 15, \\ \zeta^4 & j = 16, \ldots, 20. \end{cases}$$

(continued)

Example 7.10 (continued)

Then ϕ_1 is the identity and we can compute

$$\begin{gathered}\phi_2=\sigma,\quad \phi_3=\sigma^2,\quad \phi_4=\sigma^3,\quad \phi_5=\sigma^4,\\ \phi_6=\tau,\quad \phi_7=\sigma^3\tau,\quad \phi_8=\sigma\tau,\quad \phi_9=\sigma^4\tau,\quad \phi_{10}=\sigma^2\tau,\\ \phi_{11}=\tau^3,\quad \phi_{12}=\sigma^2\tau^3,\quad \phi_{13}=\sigma^4\tau^3,\quad \phi_{14}=\sigma\tau^3,\quad \phi_{15}=\sigma^3\tau^3,\\ \phi_{16}=\tau^2,\quad \phi_{17}=\sigma^4\tau^2,\quad \phi_{18}=\sigma^3\tau^2,\quad \phi_{19}=\sigma^2\tau^2,\quad \phi_{20}=\sigma\tau^2.\end{gathered}$$

The subgroups of F_{20} are

$$\begin{gathered}D_{10}=\langle\sigma,\tau^2\rangle,\quad C_5=\langle\sigma\rangle,\\ C_4^{(1)}=\langle\tau\rangle,\quad C_4^{(2)}=\langle\sigma^2\tau\rangle,\quad C_4^{(3)}=\langle\sigma^4\tau\rangle,\\ C_4^{(4)}=\langle\sigma\tau\rangle,\quad C_4^{(5)}=\langle\sigma^3\tau\rangle,\\ C_2^{(1)}=\langle\tau^2\rangle,\quad C_2^{(2)}=\langle\sigma^3\tau^2\rangle,\quad C_2^{(3)}=\langle\sigma\tau^2\rangle,\\ C_2^{(4)}=\langle\sigma^4\tau^2\rangle,\quad C_2^{(5)}=\langle\sigma^2\tau^2\rangle.\end{gathered}$$

Labelled in this way, $C_2^{(j)}$ and $C_4^{(j)}$ fix x_j, for $j=1,\ldots 5$. Note that τ^2 is the automorphism ϕ_{16} which interchanges ζ and ζ^4 and hence fixes $\zeta+\zeta^4$ (or equivalently $\zeta^2+\zeta^3$), which is also fixed by $\sigma=\phi_2$. Now

$$\zeta+\zeta^4=\zeta+\overline{\zeta}=2\operatorname{Re}\zeta=\tfrac{1}{2}(-1+\sqrt{5})$$

and hence D_{10} fixes $\sqrt{5}$. Since D_{10} has index 2 as a subgroup of F_{20} and the minimum polynomial of $\sqrt{5}$ has order 2, the fixed field of D_{10} is $\mathbb{Q}(\sqrt{5})$ which is the same thing as $\mathbb{Q}(\delta)$. The lattice of subgroups is

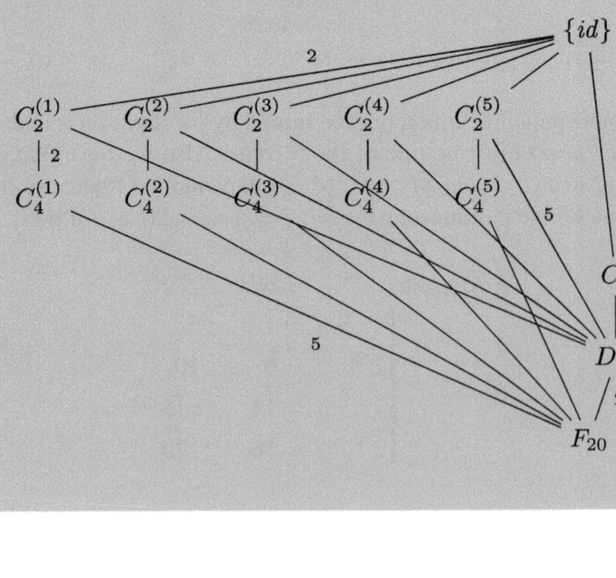

(continued)

Example 7.10 (continued)
and the corresponding subfields of $\mathbb{Q}(\alpha, \zeta)$ are

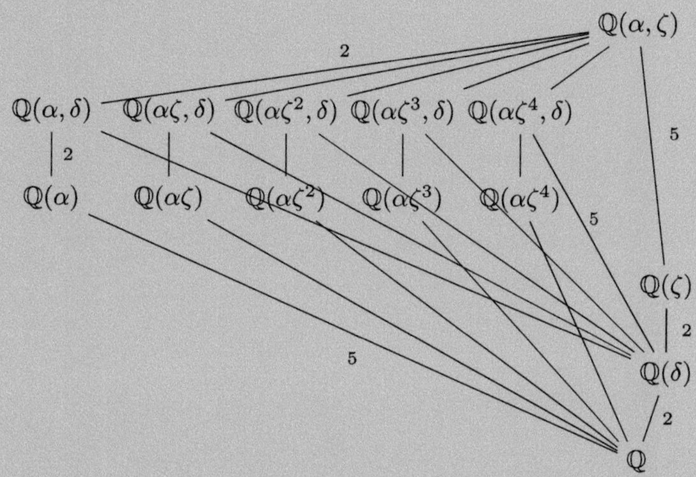

The solvability of f is a consequence of the existence of the composition series

$$F_{20} \rhd D_{10} \rhd C_5 \rhd \{id\},$$

with factors isomorphic to C_2 (twice), corresponding to the nested square roots in ζ, and C_5, corresponding to the fifth root of 2 in the solutions. ♦

8 Tschirnhaus Transformations

When it comes to quintics, the calculations can become incredibly arduous and so one avenue to explore is to look at how we can transform a general quintic into one that has some of its terms missing (just as we used the depressed cubic when deriving Cardano's formula). To that end we will look at some of the ideas of the seventeenth-century German mathematician Tschirnhaus, who, interestingly, played a leading role in the early manufacture of European porcelain. We will also cheat! Whereas Tschirnhaus and his contemporaries (and indeed all those that followed until the latter half of the twentieth century) had to do calculations by hand, or employ someone to do the calculations for them, we will make use of computer algebra software (in our case Wolfram Mathematica) to do the heavy lifting. One can only sit back and admire the tenacity of those that had to investigate the algebra of quintic equations by hand.

8.1 Tschirnhaus' Idea

In 1683 Tschirnhaus published a method which he believed could be used to solve equations of any degree by radicals. Leibniz quickly spotted the flaw in the argument; essentially that to solve an equation of degree n using Tschirnhaus' technique requires as an intermediate step the solution of an equation of degree $(n-1)!$. However, Tschirnhaus transformations can still be used to good effect to remove terms from polynomials.

Suppose

$$f(x) = x^n + a_{n-1}x^{n-1} + \cdots + a_1 x + a_0$$

C. M. Linton, *Quintic Equations and How to Solve Them*,
https://doi.org/10.1007/978-3-032-01658-4_8

has roots $x_1, \dots, x_n$. Furthermore, let $\varphi(x)$ be a polynomial with degree less than n, set $y_i = \varphi(x_i)$ and suppose that

$$g(y) = y^n + b_{n-1}y^{n-1} + \cdots + b_1 y + b_0$$

is the monic polynomial with roots y_i, $i = 1, \dots, n$. The idea is to choose φ so that some of the coefficients in the polynomial g vanish. (If we could make all of them vanish except b_0 we would have solved the polynomial; this was Tschirnhaus' goal.)

In fact, the transformations that we have already used to reduce cubics and quartics to their depressed forms are just speical cases of transformations of this type. We can take $\varphi(x) = x + \beta$ and try and eliminate b_{n-1}. Since the sums of the roots satisfy

$$x_1 + \cdots + x_n = -a_{n-1}, \quad y_1 + \cdots + y_n = -b_{n-1},$$

we require

$$\begin{aligned} 0 = y_1 + \cdots + y_n &= (x_1 + \beta) + \cdots + (x_n + \beta) \\ &= x_1 + \cdots + x_n + n\beta = -a_{n-1} + n\beta. \end{aligned}$$

Hence we should choose $\beta = a_{n-1}/n$. Thus, the cubic $x^3+bx^2+\cdots$ is transformed into its depressed form via $y = x + b/3$ and the quartic $x^4 + bx^3 + \cdots$ via $y = x + b/4$.

For the general quintic

$$x^5 + bx^4 + cx^3 + dx^2 + ex + f,$$

with roots $x_1, \dots, x_5$, this means that we should use the transformation $y = x+b/5$ to produce a quintic with no x^4 term. In fact we obtain

$$y^5 + c'y^3 + d'y^2 + e'y + f', \tag{8.1}$$

where

$$\begin{gathered} c' = c - \frac{2b^2}{5}, \quad d' = \frac{4b^3}{25} - \frac{3bc}{5} + d, \\ e' = -\frac{3b^4}{125} + \frac{3b^2c}{25} - \frac{2bd}{5} + e, \quad f' = \frac{4b^5}{3125} - \frac{b^3c}{125} + \frac{b^2d}{25} - \frac{be}{5} + f, \end{gathered}$$

with roots $y_i = x_i + b/5$. In this chapter we will use the notation $s_j^{(x)}$ for the elementary symmetric polynomials in the roots $x_1, \dots, x_5$ and $s_j^{(y)}$ for the elementary symmetric polynomials in the roots $y_1, \ldots, y_5$. Hence

$$s_1^{(x)} = -b, \quad s_2^{(x)} = c, \quad s_3^{(x)} = -d, \quad s_4^{(x)} = e, \quad s_5^{(x)} = -f,$$
$$s_1^{(y)} = 0, \quad s_2^{(y)} = c', \quad s_3^{(y)} = -d', \quad s_4^{(y)} = e', \quad s_5^{(y)} = -f'.$$

The calculations below will make extensive use of the sums of the powers of the roots and so we define

$$P_m^{(x)} = x_1^m + x_2^m + x_3^m + x_4^m + x_5^m,$$

with a similar definition for $P_m^{(y)}$ in terms of the roots $y_1, \ldots, y_5$. Expressions for these sums in terms of the elementary symmetric polynomials, and hence in terms of the coefficients in the quintic, can be obtained using identities that are usually named after Newton, but were implicit in the work of Girard, writing over 30 year before Newton formulated them; see [13, p. 38]. The Newton-Girard identities are, for quintic polynomials, omitting the superscript (x) or (y) as they can be applied to either case,

$$P_2 = s_1 P_1 - 2s_2,$$
$$P_3 = s_1 P_2 - s_2 P_1 + 3s_3,$$
$$P_4 = s_1 P_3 - s_2 P_2 + s_3 P_1 - 4s_4,$$
$$P_5 = s_1 P_4 - s_2 P_3 + s_3 P_2 - s_4 P_1 + 5s_5,$$
$$P_m = s_1 P_{m-1} - s_2 P_{m-2} + s_3 P_{m-3} - s_4 P_{m-4} + s_5 P_{m-5}, \quad m > 5.$$

Starting from $P_1^{(x)} = s_1^{(x)} = -b$, we can calculate $P_2^{(x)}, P_3^{(x)}, \ldots$ in turn. We obtain

$$P_2^{(x)} = b^2 - 2c, \qquad P_3^{(x)} = -b^3 + 3bc - 3d,$$
$$P_4^{(x)} = b^4 - 4b^2c + 4bd + 2c^2 - 4e.$$

Expressions for $P_m^{(y)}$ are identical in terms of the dashed coefficients, except that $b' = 0$.

As we have seen, a linear transformation is all that is required to remove the x^4 term and it is relatively straightforward to extend this and use a quadratic Tschirnhaus transformation $y = \varphi(x) = x^2 + \beta x + \gamma$ to reduce the general quintic to one with roots $y_i = \varphi(x_i)$ in which there is no quartic or cubic term:

$$g(y) = y^5 + py^2 + qy + r.$$

This is known as a **principal quintic**, though also sometimes referred to as a canonical quintic, as in [69], which explores the geometry which lies behind Tschirnhaus transformations.

To see how we can first determine β and γ we consider the sum of the roots $y_1 + \cdots + y_5$ which must be zero:

$$0 = P_1^{(y)} = P_2^{(x)} + \beta P_1^{(x)} + 5\gamma = b^2 - 2c - \beta b + 5\gamma,$$

and so

$$5\gamma = \beta b + 2c - b^2. \tag{8.2}$$

Similarly, by considering the sum $P_2^{(y)} = y_1^2 + \cdots + y_5^2$ we can show that

$$(2b^2-5c)\beta^2+(-4b^3+13bc-15d)\beta+2b^4-8b^2c+10bd+3c^2-10e = 0, \tag{8.3}$$

which is a quadratic equation for β, unless $2b^2 = 5c$. If $2b^2 = 5c$ then, from (8.1), there is no need to apply the quadratic transformation since the linear transformation $y = x+b/5$ already reduces the given quintic to a principal one. Provided β satisfies this quadratic equation, with γ chosen according to (8.2), the resulting polynomial for y will have no third or fourth powers; hence we can take either root for β. It is worth noting first that β and γ do not depend on the constant term f and that in applying this transformation we are potentially introducing a square root (of the discriminant of the quadratic for β).

The more difficult part of the process is the determination of the coefficients p, q and r that result if we apply this quadratic transformation and there is no getting around the fact that the algebra involved is laborious. We will spare the reader any of the details; a particularly neat way of expressing the result is that p, q and r can be determined in turn from

$$\begin{aligned} 10\gamma^3 + p &= d\beta_3 + e(4\beta^2 - 3b\beta + 2c) + f(5\beta - 2b), \\ 5\gamma^4 + 2p\gamma + q &= e\beta_4 + f(5\beta^3 - 4b\beta^2 + 3c\beta - 2d), \\ \gamma^5 + p\gamma^2 + q\gamma + r &= f\beta_5, \end{aligned} \tag{8.4}$$

where

$$\beta_3 = \beta^3 - b\beta^2 + c\beta - d, \quad \beta_4 = \beta^4 - b\beta^3 + c\beta^2 - d\beta + e,$$
$$\beta_5 = \beta^5 - b\beta^4 + c\beta^3 - d\beta^2 + e\beta - f.$$

The eagle-eyed reader with a knowledge of basic calculus will notice that the expressions on the left-hand sides of the equations in (8.4) are, in ascending order, $g(\gamma)$, $g'(\gamma)$ and $g''(\gamma)$, which gives a hint at how these equations are derived.

Example 8.1 Consider the quintic

$$\chi(x) = x^5 + x^4 + x^3 + x^2 + x + f,$$

with roots $x_1, \ldots, x_5$. The equations (8.2) and (8.3) in this case reduce to $5\gamma = \beta + 1$ and $(\beta + 1)^2 = 0$ and so we can reduce χ to a principal quintic by a quadratic transformation of the form $y = x^2 + \beta x + \gamma$ with $\beta = -1$ and $\gamma = 0$, i.e. $y = x^2 - x$. This can then be shown, from (8.4), to lead to

$$y^5 + (5 - 7f)y^2 + (5 - 14f)y - f(5 + f) \tag{8.5}$$

with roots $y_i = x_i^2 - x_i$, $i = 1, \ldots, 5$. ...

It would appear that in order to determine the roots x_i in the above example, assuming we can find the roots y_i, we need to solve a quadratic, only one of whose roots will be a root to the original quintic, which is somewhat unsatisfactory. There is an exception to this, namely if the principal quintic has a double root y^*, since then both roots of the quadratic $x^2 - x - y^*$ will be roots of our quintic. However, there is a better way.

If we take the equation which defines the transformation and successively multiply by x, we get, writing $\tilde{y} = y - \gamma$,

$$\tilde{y} = x^2 + \beta x, \quad \tilde{y}x = x^3 + \beta x^2, \quad \tilde{y}x^2 = x^4 + \beta x^3.$$

We can then express powers of x as linear functions of x:

$$\begin{aligned}
x^2 &= \tilde{y} - \beta x, \\
x^3 &= (\tilde{y} + \beta^2)x - \beta\tilde{y}, \\
x^4 &= \left(-\beta^3 - 2\beta\tilde{y}\right)x + \tilde{y}^2 + \beta^2\tilde{y}.
\end{aligned}$$

But we also have

$$\tilde{y}x^3 = x^5 + \beta x^4 = (\beta - b)x^4 - cx^3 - dx^2 - ex - f,$$

where we have used the original quintic to replace the x^5 term. We can now substitute all the powers of x in this equation with linear functions of x and rearrange to yield

$$\left(\tilde{y}^2 + (c - 2b\beta + 3\beta^2)\tilde{y} + \beta_4\right)x = (2\beta - b)\tilde{y}^2 + \beta_3\tilde{y} - f. \tag{8.6}$$

Provided the factor on the left-hand side is non-zero, this can be used to determine the roots x_i unambiguously once the roots y_i are known.

...Continuation of Example 8.1. From (8.6), the roots of the principal quintic (8.5) are related to the roots of $x^5 + x^4 + x^3 + x^2 + x + f$ via

$$(y_i + 5)(y_i + 1)x_i = -(3y_i^2 + 4y_i + f), \quad i = 1, \ldots, 5, \tag{8.7}$$

which allows us to determine x_i provided y_i is not -1 or -5.

In the special case when $f = 1$, so that the roots for x are the five 6th roots of unity other than 1, namely $-1, \pm\omega, \pm\omega^2$, the associated principal quintic is

$$y^5 - 2y^2 - 9y - 6 = (y + 1)^2(y - 2)(y^2 + 3)$$

with roots $-1, -1, 2, \pm i\sqrt{3}$. The relationship (8.7) between the roots of the quintics for x and y holds for $y = 2$ and $y = \pm i\sqrt{3}$, leading to $x = -1, \omega, \omega^2$. However, we cannot use (8.7) for the double root at $y = -1$ since it just says $0 = 0$. Instead we can solve the quadratic $x^2 - x = -1$ to yield the other two roots $-\omega, -\omega^2$. This also implies that $x^2 - x + 1$ is a factor of our original quintic and indeed we have

$$x^5 + x^4 + x^3 + x^2 + x + 1 = (x^2 - x + 1)(x^2 + x + 1)(x + 1).$$

The form of (8.7) suggests that there will be another quintic of the form $x^5 + x^4 + x^3 + x^2 + x + f$ for which the associated principal quintic has a double root. We require that the expression $3y^2 + 4y + f$ has a factor of $y + 5$ and this happens if $f = -55$. For this case we find the principal quintic

$$y^5 + 390y^2 + 775y - 2750 = (y + 5)^2(y^3 - 10y^2 + 75y - 110).$$

The roots for x that correspond to the double root $y = -5$ are the roots of the quadratic $x^2 - x + 5$, which must be a factor of our original quintic. In fact

$$x^5 + x^4 + x^3 + x^2 + x - 55 = (x^2 - x + 5)(x^3 + 2x^2 - 2x - 11),$$

both factors on the right-hand side being irreducible over $\mathbb{Q}$. ♦

The problem of finding the roots of a quintic can thus always be reduced, by either a linear or a quadratic Tschirnhaus transformation, to that of finding the roots of a principal quintic. As Tschirnhaus was aware, the method outlined above can be used to remove the second and third terms from a polynomial of any degree and so, in particular, it can be used to solve the general cubic, by reducing it to the form $x^3 + d$. For the quartic, Lagrange showed that a quadratic Tschirnhaus transformation can be used to reduce a general quartic to a biquadratic (with no x^3 or x terms) which is then easily solved. This requires the solution of a cubic equation along the way. For polynomials of higher degree than three, the removal of the second, third and fourth terms using a cubic transformation requires the solution of an equation of degree 6, but in the specific case of the quartic (and only for the quartic), Lagrange showed that this sextic factorises into three quadratics, each of which has coefficients that are solutions to a cubic equation and hence a cubic Tschirhnaus transformation can be used to reduce a general quartic to the form $x^4 + e$. However, these approaches are cumbersome and add little to what we have already discussed and so we will not dwell on them.

8.2 Bring and Jerrard

Tschirnhaus appears to have believed that a cubic transformation could be used to reduce the general quintic to one of the form

$$y^5 + qy + r. \tag{8.8}$$

However, as we have noted, this requires the solution of a sextic equation. The first person to get round this problem was the amateur Swedish mathematician Erland Bring in 1786, though his work seems not to have become widely known and was only rediscovered when George Jerrard proposed an alternative approach in 1834, apparently unaware of Bring's work; see [22, pp. 5–7] and [23]. Jerrard's technique is more general in that it allows one to remove the second, third and fourth terms from a general polynomial equation of any order greater than four. In fact, for the general quintic equation, Jerrard's method can be used to produce any of the forms

$$y^5 + qy^j + r, \quad j = 1, 2, 3, 4. \tag{8.9}$$

A quintic of the form (8.8) is known as a Bring-Jerrard quintic.

We will describe Jerrard's approach as it has a slightly different flavour to the way Bring approached the problem, the latter being more akin to the original approach of Tschirnhaus. Jerrard actually claimed his method would allow a general quintic to be reduced to a solvable form but William Rowan Hamilton, in a detailed exposition of the power and limitations of Jerrard's approach, showed this to be false. The key to both Bring's and Jerrard's methods is to introduce more parameters and thus instead of a cubic transformation, we use a quartic.

To keep the algebra manageable (just about!) we start from the principal quintic

$$x^5 + dx^2 + ex + f, \quad d \neq 0$$

and attempt to reduce this to the Bring-Jerrard quintic

$$y^5 + qy + r$$

with the aid of the transformation

$$y = \alpha_4 x^4 + \alpha_3 x^3 + \alpha_2 x^2 + \alpha_1 x + \alpha_0. \tag{8.10}$$

Jerrard kept the coefficient of x^4 in this expression free rather than setting it to one, for reasons that will become apparent shortly. The first observation is that y^5 involves powers of x up to 20 and so we will need to calculate $P_m^{(x)}$ for all $m \leq 20$. This is a tedious task, though given that we are working with a principal quintic, with no x^4 or x^3 terms, the expressions remain fairly compact. After $P_1^{(x)} = P_2^{(x)} = 0$ the next few are:

$$P_3^{(x)} = -3d, \quad P_4^{(x)} = -4e, \quad P_5^{(x)} = -5f,$$
$$P_6^{(x)} = 3d^2, \quad P_7^{(x)} = 7de, \quad P_8^{(x)} = 4(e^2 + 2df).$$

In order that the quintic with roots $y_1, \ldots, y_5$ has no quartic, cubic or quadratic terms, we require that the first three elementary symmetric polynomials are zero, which in turn implies that

$$P_1^{(y)} = P_2^{(y)} = P_3^{(y)} = 0.$$

Consideration of $P_1^{(y)}$ gives

$$0 = P_1^{(y)} = \alpha_4 P_4^{(x)} + \alpha_3 P_3^{(x)} + 5\alpha_0.$$

Hence we take

$$\alpha_0 = \tfrac{1}{5}(3d\alpha_3 + 4e\alpha_4).$$

To satisfy $P_2^{(y)} = 0$ we need to square (8.10) and sum over the roots $x_1, \ldots, x_5$. This leads to the requirement that

$$2(e^2 + 10df)\alpha_4^2 + 3d^2\alpha_3^2 - 10e\alpha_2^2 + 23de\alpha_3\alpha_4$$
$$+ 5\alpha_2(3d^2\alpha_4 - 5f\alpha_3) - 5\alpha_1(3d\alpha_2 + 4e\alpha_3 + 5f\alpha_4) = 0. \tag{8.11}$$

The crucial thing about this expression for our purposes is that it is **homogeneous of degree 2** in the variables $\alpha_1, \alpha_2, \alpha_3, \alpha_4$. In other words each term contains two of the quantities $\alpha_1, \alpha_2, \alpha_3, \alpha_4$ (which may be the same) multiplied together. This is the reason that we left α_4 as a parameter, rather than setting it to one.

Now for Jerrard's clever idea. We write the left-hand side of (8.11) in the form

$$(l_1\alpha_3 + l_2\alpha_4 + l_3\alpha_1 + l_4\alpha_2)^2 + (l_5\alpha_4 + l_6\alpha_1 + l_7\alpha_2)^2 + (l_8\alpha_1 + l_9\alpha_2)^2 + l_{10}\alpha_2^2.$$

The reason we single out α_3 as the only coefficient that appears in just one of the four factors is that the coefficient of α_3^2 in (8.11) is $3d^2$ which is non-zero by assumption; the coefficients of α_1^2, α_2^2 and α_4^2 either are, or could be, zero, which would cause a problem. If we expand this out and compare with the left-hand side of (8.11) we see that

$$l_1^2 = 3d^2, \quad 2l_1l_2 = 23de, \quad l_1l_3 = -10e, \quad 2l_1l_4 = -25f,$$
$$l_2^2 + l_5^2 = 2(e^2 + 10df), \quad 2(l_2l_3 + l_5l_6) = -25f,$$
$$2(l_2l_4 + l_5l_7) = 15d^2, \quad l_3^2 + l_6^2 + l_8^2 = 0,$$
$$2(l_3l_4 + l_6l_7 + l_8l_9) = -15d, \quad l_4^2 + l_7^2 + l_9^2 + l_{10}^2 = -10e$$

and these equations allow us to calculate $l_1, l_2, \ldots, l_{10}$ in turn. We can then satisfy (8.11) by choosing $\alpha_1, \alpha_2, \alpha_3, \alpha_4$ so that

$$l_1\alpha_3 + l_2\alpha_4 + l_3\alpha_1 + l_4\alpha_2 = \mathrm{i}(l_5\alpha_4 + l_6\alpha_1 + l_7\alpha_2), \quad l_8\alpha_1 + l_9\alpha_2 = \mathrm{i}l_{10}\alpha_2.$$

The second of these equations gives α_1 as a multiple of α_2 and then the first equation gives α_3 as a linear combination of α_2 and α_4. If we write

$$\alpha_1 = A\alpha_2, \quad \alpha_3 = B\alpha_2 + C\alpha_4,$$

with

$$A = (\mathrm{i}l_{10} - l_9)/l_8, \quad B = (\mathrm{i}l_6 - l_3)A/l_1 + (\mathrm{i}l_7 - l_4)/l_1, \quad C = (\mathrm{i}l_5 - l_2)/l_1,$$

then (8.11) holds. Jerrard's idea is a special case of the theorem [2, Prop. 10.1] that if $V(\alpha_1, \ldots, \alpha_n)$ is a homogeneous function of degree 2 in the variables $\alpha_1, \ldots, \alpha_n$, then V can be expressed as the sum of n squares.

We now have one more condition to satisfy, namely $P_3^{(y)} = 0$, which leads to a cubic in α_2/α_4 (far too complicated to write down here) and then q and r follow from $P_4^{(y)} = -4q$ and $P_5^{(y)} = -5r$. If we want to produce a form with an x^2 term but no x term, then instead we need to satisfy $P_4^{(y)} = 0$, which leads to a quartic for α_2/α_4. The other two forms in (8.9) can be obtained by working with the reciprocal quintic (in powers of $1/x$).

Finally, when we have calculated the roots y_k we can then determine the roots x_k as a rational function of y_k by taking

$$y = \alpha_4x^4 + \alpha_3x^3 + \alpha_2x^2 + \alpha_1x + \alpha_0$$

and deriving expressions for yx, yx^2, yx^3, yx^4 as linear combinations of $1, x, x^2, x^3, x^4$, replacing x^5 by $-dx^2 - ex - f$ whenever it occurs. For example, we will end up with an equation of the form

$$yx^4 = c_4x^4 + c_3x^3 + c_2x^2 + c_1x + c_0,$$

for some set of coefficients $c_0, \ldots, c_4$, which we can solve for x^4 in terms of x^3, x^2, x and y. We will have a similar equation for yx^3 that can now be solved for x^3 in terms of x^2, x and y an we can repeat the process until we have an expression for x as a rational function of y.

We have thus shown that we can always reduce the problem of solving a general quintic to one of finding the roots of a quintic of the form $y^5 + py + q$, through a process which involves solving nothing worse than a cubic equation. This is of theoretical interest, though of no real practical value. The calculations involved are horrendous and, even for specific numerical values of the coefficients, there is no way to use this technique to actually find exact expressions for p and q in terms of d, e and f. On the other hand, at the time that these techniques were being developed, when the mysteries of the quintic were only just being unearthed, it was clearly important to be able to reduce the number of parameters that needed to be considered and Bring and Jerrard both showed how this could be done using radicals alone. Note that although the Bring-Jerrard quintic has two parameters, it is easily reduced to one with a single parameter. If we write $y = qz/p$, then $y^5 + py + q = 0$ becomes

$$z^5 + az + a = 0,$$

where $a = p^5/q^4$.

Solvable Quintics and How to Solve Them 9

Towards the end of the nineteenth century two men, little known in the history of mathematics, developed approaches which enable one to determine whether a given quintic is solvable by radicals and if so, how to accomplish the task. The first was George Paxton Young, who considered quintics with the special form $x^5 + ex + f$ (known as Bring-Jerrard quintics, see Sect. 8.2) and then both he and Emory McClintock devised approaches for the general quintic. Others also worked on these problems, but it was Young and McClintock who published clear expositions of their methods and interestingly, neither of them was a professional mathematician. Young and McClintock each published three notable papers related to the solution of the quintic [72–74] and [46–48], respectively.

Young, who was born in 1818, was a clergyman, educated at the University of Edinburgh, who emigrated to Canada in 1850 and was appointed to the chair of metaphysics and ethics at University College, Toronto in 1871. He was an educator with very broad interests who wrote on many topics, including philosophy, theology and education policy as well as the subject that was something of a hobby to him—mathematics. McClintock was born in Pennsylvania in 1840 and studied in both the USA and Europe. On returning from Paris to America in 1867 he became an actuary and this was the profession that he practiced for the rest of his life. At the time of his retirement he was Vice President of the Mutual Life Insurance Company in New York and one of the leading actuaries in the United States. McClintock was very active in mathematical circles and served as President of the American Mathematical Society between 1891 and 1894.

Of the two, Young's work on the quintic appears to be better known. He produced a simple characterisation of Bring-Jerrard quintics that can be solved by radicals and a fairly simple method for then solving them. He also developed a general method for solving solvable quintics, though how he did this was less clear. George Neville Watson, who became interested in quintic equations through his attempts to verify results published by Ramanujan, later commented that "his results seemed to be obtained by a sheer piece of good fortune, and not as a consequence of deliberate and

C. M. Linton, *Quintic Equations and How to Solve Them*,
https://doi.org/10.1007/978-3-032-01658-4_9

systematic strategy" [4]. Watson gave a lecture in 1948 in which he proudly claimed to have solved over 100 quintics using his own method, which was essentially an improved version of Young's (interestingly the notes for this lecture remained in a box for about 50 years before being rediscovered in the 1990s). Watson did not mention the work of McClintock even though McClintock's final paper on the topic in 1898 goes even further than Watson did and so it seems likely he was unaware of it. In this Chapter we will describe how to identify and solve solvable quintics in a way which combines the contributions of all of these authors. A succinct summary of the theory of quintic equations as it stood in the 1920s, including references to the primary sources, is given in [17], though only two brief paragraphs (§112) are given to the subject of solvable quintics.

9.1 Basic Properties of Quintics

For simplicity, we will only consider quintics with rational coefficients. Let

$$q(x) = x^5 + cx^3 + dx^2 + ex + f \tag{9.1}$$

be a polynomial with $c, d, e, f \in \mathbb{Q}$ and $f \neq 0$. (If f is zero, then one of the roots is $x = 0$ and the others are the roots of a quartic.) The roots of q will be denoted by $x_1, \ldots, x_5$ and none of the roots is zero. Many of the formulas that we will encounter are more easily expressed in terms of coefficients that are scaled using the fifth-power binomial expansion, i.e. 1, 5, 10, 10, 5, 1, so we will also write

$$q(x) = x^5 + 10Cx^3 + 10Dx^2 + 5Ex + f, \tag{9.2}$$

where $c = 10C$, $d = 10D$ and $e = 5E$. These two forms will be used interchangeably. We will refer to (9.1) or (9.2) as a general quintic even though there is no x^4 term. If we have a quintic with a non-zero x^4 term then, as we saw in the previous chapter, it can be turned it into the form (9.1) using a simple translation. For the quintic $y^5 + by^4 + \ldots$ then the substitution $y = x + b/5$ will yield a quintic for x with no x^4 term.

The discriminant of the depressed quintic (9.1) can be computed to be

$$\begin{aligned}\Delta = {}& -4c^3d^2e^2 - 27d^4e^2 + 16c^4e^3 + 144cd^2e^3 - 128c^2e^4 + 256e^5\\ &+ 16c^3d^3f + 108d^5f - 72c^4def - 630cd^3ef + 560c^2de^2f\\ &- 1600de^3f + 108c^5f^2 + 825c^2d^2f^2 - 900c^3ef^2\\ &+ 2250d^2ef^2 + 2000ce^2f^2 - 3750cdf^3 + 3125f^4\end{aligned} \tag{9.3}$$

and for a quintic in Bring-Jerrard form this reduces to

$$\Delta = 256e^5 + 3125f^4 = 4^4e^5 + 5^5f^4. \tag{9.4}$$

Unless stated otherwise, we will assume that the quintic (9.1) is irreducible and hence (see Proposition 3.1) separable, which implies that $\Delta \neq 0$. The square root of the discriminant will be denoted by δ (where it matters which square root we choose, we will say so).

We saw in Sect. 7.4 that the Galois group of an irreducible quintic is either S_5, A_5, F_{20}, D_{10} or C_5. For the latter three of these, which are the possible Galois groups of solvable quintics, there are 6 conjugate subgroups which correspond to different orderings of the roots and so we will make precise which particular subgroup we are dealing with. This will facilitate calculations. As previously, we will take

$$\sigma = (x_1x_2x_3x_4x_5) \quad \text{and} \quad \tau = (x_2x_3x_5x_4)$$

and with these choices the elements of F_{20} are listed explicitly in Sect. 7.4. Choosing a specific version of F_{20} has the effect of restricting the possible ordering of the roots $x_1, \ldots, x_5$. We then take D_{10} to be the intersection of A_5 with the particular version of F_{20} generated by σ and τ and C_5 to be the cyclic group generated by $\sigma = (x_1x_2x_3x_4x_5)$, which is then a subgroup of D_{10}.

We noted in Proposition 7.3 that solvable, irreducible quintics have either one or five real roots and it can be shown that for five real roots we require each of the four following inequalities to be satisfied:

$$c < 0, \quad D_3 > 0, \quad D_4 > 0, \quad D_5 > 0. \tag{9.5}$$

where

$$\begin{aligned}
D_3 &= 40ce - 12c^3 - 45d^2,\\
D_4 &= 12c^4e - 4c^3d^2 + 117cd^2e - 88c^2e^2 - 40c^2df + 125cf^2\\
&\quad - 27d^4 - 300def + 160e^3,\\
D_5 &= -1600de^3f - 3750cdf^3 + 2000ce^2f^2 - 4c^3d^2e^2 + 16c^3d^3f\\
&\quad - 900c^3ef^2 + 825c^2d^2f^2 + 144cd^2e^3 + 2250d^2ef^2 + 16c^4e^3\\
&\quad + 108c^5f^2 - 128c^2e^4 - 27d^4e^2 + 108d^5f + 256e^5 + 3125f^4\\
&\quad - 72c^4def + 560c^2de^2f - 630cd^3ef.
\end{aligned}$$

(This test is given in [71] and reproduced in [70].) Otherwise any solvable, irreducible quintic (9.1) has a single real root. In particular, any solvable, irredcucible principal quintic (i.e. with $c = 0$) has only a single real root. We have also shown previously (Proposition 7.4) that solvable, irreducible quintics with one real root have as their Galois group either F_{20} or D_{10}.

9.2 Sextic Resolvents

The key to solving solvable quintic equations, i.e. irreducible quintics with Galois group F_{20}, D_{10} or C_5, is the construction of a particular resolvent polynomial of degree 6. This resolvent was originally discovered by Carl Gustav Jacobi in 1835, rediscovered by James Cockle and systematically investigated by Arthur Cayley, who wrote 12 papers on quintic equations in 1860–1862 alone; see [15]. While Jacobi had found the form of this resolvent without computing the numerical value of the coefficients, Cayley calculated it in full.

We have seen that the Galois group of a solvable irreducible quintic must be a subgroup of F_{20} and any combination of the roots $x_1, \ldots, x_5$ which is fixed by F_{20} but not by any other elements of S_5 will take $120/20 = 6$ values under the action of S_5 and these 6 values are the roots of a sextic polynomial. The solvability or otherwise of our quintic will turn out to depend on whether or not this sextic has a rational root, something that can be checked using the rational root test (see Sect. 3.2).

We define the symbol $\oplus$ (chosen to signify a combination of addition and a cyclic permutation and no connection to its usual meaning in linear algebra as a direct sum) applied to a polynomial function p of $x_1, \ldots, x_5$ via

$$\oplus\,[p] = p + \sigma p + \sigma^2 p + \sigma^3 p + \sigma^4 p,$$

with $\sigma = (x_1 x_2 x_3 x_4 x_5)$. Note that it follows immediately from the definition that σ fixes $\oplus\,[p]$ for any p.

Consider

$$\psi = \oplus[x_1(x_2 - x_3)] \tag{9.6}$$

which according to our definition is the same as

$$\psi = x_1(x_2 - x_3) + x_2(x_3 - x_4) + x_3(x_4 - x_5) + x_4(x_5 - x_1) + x_5(x_1 - x_2).$$

For any $\mu \in S_5$, $\mu\psi$ is a sum of 10 terms, each of which is the product of two distinct roots and there are only $(5 \times 4)/2 = 10$ such terms. Hence the action of μ on ψ is merely to change the sign of some of the terms. We can illustrate the combination of signs via a pentagram, a device that appears to have been introduced by Watson; see Fig. 9.1. The points are labelled $1, 2, 3, 4, 5$ and if two numbers j and k represent adjacent vertices, then the $x_j x_k$ term in the expression for ψ is positive, whereas if they are opposite vertices, joined by a solid line, then the term has a minus sign.

The action of μ on ψ corresponds to a permutation of the numbers $\{1, 2, 3, 4, 5\}$ in the diagram. Without loss of generality, we can label a fixed vertex as 1 and

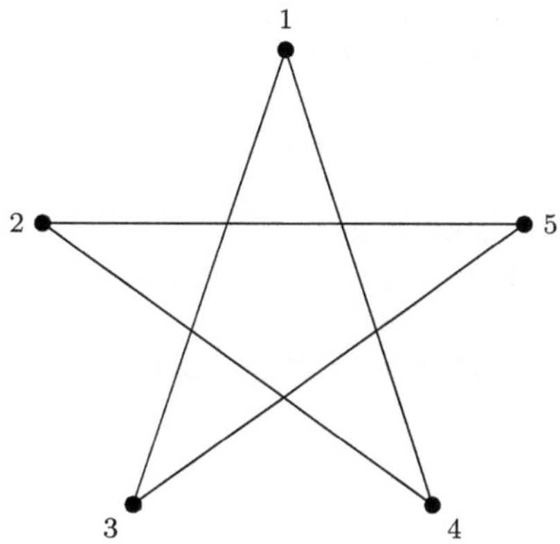

Fig. 9.1 Pentagram representing the signs of the terms in ψ. If two numbers j and k are adjacent then the $x_j x_k$ term is positive, whereas if they are opposite vertices then the term has a minus sign. Thus the arrangement shown corresponds to $x_1x_2 - x_1x_3 - x_1x_4 + x_1x_5 + x_2x_3 - x_2x_4 - x_2x_5 + x_3x_4 - x_3x_5 + x_4x_5$ as in (9.6). Permuting the labels 2, 3, 4, 5 leads to different forms for $\mu\psi$, where $\mu \in S_5$

then there are $4! = 24$ possibilities for the labels on the other vertices. If we reflect the pentagram about a line of symmetry through one vertex, this does not change the adjacencies and so there are only 12 different possibilities for $\mu\psi$. Looking at Fig. 9.1, if we interchange the upper pair of vertices on the left- and right-hand sides with the lower pair, then adjacencies become opposites and vice versa. Hence in this case, all the signs change and ψ becomes $-\psi$. It follows that ψ^2 takes only 6 values under the action of any of the 120 elements of S_5 and these 6 values will thus be the roots of a sextic polynomial, which we will now derive.

Let the 12 values taken by ψ be $\pm\psi_1, \pm\psi_2, \pm\psi_3, \pm\psi_4, \pm\psi_5, \pm\psi_6$, where $\psi_1, \ldots, \psi_6$ correspond to pentagrams with labels 12345, 13425, 14235, 12534, 13542, 14523, respectively, each associated with a different conjugate version of F_{20}. The pentagram in the figure is the one that corresponds to $\sigma = (x_1x_2x_3x_4x_5)$ and so ψ_1 is the same as ψ. If we define

$$P_m = \psi_1^m + \psi_2^m + \psi_3^m + \psi_4^m + \psi_5^m + \psi_6^m, \tag{9.7}$$

then it can be shown that any transposition takes P_m to $(-1)^m P_m$. We know that the square root of the discriminant,

$$\delta = (x_1 - x_2)(x_1 - x_3)(x_1 - x_4)(x_1 - x_5)(x_2 - x_3) \times (x_2 - x_4)(x_2 - x_5)(x_3 - x_4)(x_3 - x_5)(x_4 - x_5),$$

maps to $-\delta$ under the action of any transposition. From this it can be shown that δ, in which each term is the product of 10 roots, must be a factor of P_m whenever m is odd. However, P_1 is of degree 2 in the roots and P_3 is of degree 6 and since these are less than 10, P_1 and P_3 must be identically zero, whereas P_5 has degree 10 and therefore must be a multiple of δ; in fact we can show that $P_5 = 160\delta$. On the other and, if m is even, P_m is a symmetric polynomial in $x_1, \ldots, x_5$ and hence can be expressed as a polynomial function of the coefficients of the quintic. We conclude that the sextic (the Jacobi-Cayley resolvent)

$$J(x) = (x - \psi_1)(x - \psi_2)(x - \psi_3)(x - \psi_4)(x - \psi_5)(x - \psi_6)$$

has coefficients in $\mathbb{Q}(\delta)$.

If we write $s_1, \ldots, s_6$ for the elementary symmetric polynomials in $\psi_1, \ldots, \psi_6$, then from the Newton-Girard identities for sextic polynomials [13, p. 38] we have $s_1 = P_1 = s_3 = P_3 = 0$ and

$$\begin{aligned} P_2 &= -2s_2, \quad & P_4 &= -s_2 P_2 - 4s_4, \\ P_5 &= 5s_5 = 160\delta, \quad & P_6 &= -s_2 P_4 - s_4 P_2 - 6s_6. \end{aligned}$$

We can then express s_2, s_4 and s_6 (and hence the coefficients in $J(x)$) in terms of C, D, E, f. This involves reducing the expressions for P_2, P_4 and P_6 in terms of $x_1, \ldots x_5$ to combinations of symmetric polynomials in the roots of the quintic. This would be an extremely laborious task without the help of a computer algebra package. It turns out to be convenient to work with the scaled quantities S_2, S_4 and S_6 defined by

$$s_2 = 100S_2, \quad s_4 = 2000S_4, \quad s_6 = 40{,}000S_6$$

and then we find that

$$\begin{aligned} S_2 &= -3C^2 - E, \\ S_4 &= 15C^4 + 8CD^2 - 2C^2E + 3E^2 - 2Df, \\ S_6 &= -25C^6 - 40C^3D^2 + 35C^4E - 11C^2E^2 - 2C^2Df - 16D^4 \\ &\qquad + E^3 - 2DEf + 28CD^2E + Cf^2. \end{aligned} \tag{9.8}$$

With these values, the Jacobi-Cayley resolvent is

$$J(x) = x^6 + s_2x^4 + s_4x^2 - 32\delta x + s_6, \tag{9.9}$$

with roots $\psi_1, \ldots, \psi_6$. The quantity $s_6 = \psi_1\psi_2\psi_3\psi_4\psi_5\psi_6$ plays an important role in what follows since $s_6 = 0$ is the condition for zero to be a root of the resolvent sextic. Here we note the alternative form

$$CS_6 = (C^2D + DE - Cf)^2 - (C^3 + D^2 - CE)((5C^2 - E)^2 + 16CD^2) \tag{9.10}$$

which can be used when $C \neq 0$. If we set $J(x)$ to zero, take the δx term to the other side of the equation and square the result, we obtain

$$(y^3 + s_2y^2 + s_4y + s_6)^2 = 2^{10}\Delta y, \tag{9.11}$$

where we have written $y = x^2$. This is a sextic for y with rational coefficients and roots $\psi_1^2, \ldots, \psi_6^2$.

We now appeal to Galois theory to provide a test for the solvability of our quintic. The particular conjugate subgroup F_{20} that we are using, generated by $\sigma = (x_1x_2x_3x_4x_5)$ and $\tau = (x_2x_3x_5x_4)$, includes precisely those elements of S_5 which fix ψ_1^2; hence ψ_1^2 is an element of the fixed field of F_{20}. It follows from the inclusion reversing nature of the Galois correspondence (see Sect. 6.4), that since the Galois group G_q of any solvable irreducible quintic must be contained within F_{20}, the fixed field of F_{20} is contained within the fixed field of G_q, which is $\mathbb{Q}$. It also contains $\mathbb{Q}$ and hence the fixed field of F_{20}, acting on the splitting field of a solvable irreducible quintic, is precisely $\mathbb{Q}$. Our quintic will thus be solvable if and only if ψ_1^2 is rational. It can also be shown that if this sextic has a rational root, then it factors into a linear polynomial and an irreducible quintic; see [19, Thm. 1].

If ψ is a root of $J(x)$ for which ψ^2 is rational, then it is clear from (9.9) that $\delta\psi$ must also be rational. In other words, $\psi = \rho_S\delta$ for some rational ρ_S. If we set $x = \rho\delta$ in the expression for $J(x)$, we find that the sextic with root $\rho = \rho_S$ is

$$\Delta^3\rho^6 + s_2\Delta^2\rho^4 + s_4\Delta\rho^2 - 32\Delta\rho + s_6, \tag{9.12}$$

with rational coefficients. Finding rational roots to this sextic is a matter of applying the rational root test which, although it may be laborious, is a finite process. This sextic polynomial is easier to deal with than (9.11) as it has simpler coefficients and no quintic or cubic terms.

We can perform the same analysis for the resolvent

$$\Psi = \oplus[x_1x_2(x_3 - x_4)], \tag{9.13}$$

which takes 12 values under the action of S_5 in exactly the same way as ψ given by (9.6). For this case, we define $\Psi_1, \ldots \Psi_6$ using the same 6 pentagrams as for $\psi_1, \ldots \psi_6$, but now the sign of the term $x_ix_jx_k$ is determined from the adjacency or otherwise of the two roots *not* present in a given term. If we define P_m as in (9.7) but with ψ_k replaced by Ψ_k, then we again conclude that $P_1 = P_3 = 0$ (since they have degrees 3 and 9, respectively, which are less than 10) and that δ is a factor of P_5. In this case P_5 is of degree 15 and so there is an additional factor of degree 5, which must be a symmetric polynomial and hence rational. Thus

$$K(x) = (x - \Psi_1)(x - \Psi_2)(x - \Psi_3)(x - \Psi_4)(x - \Psi_5)(x - \Psi_6)$$

has coefficients in $\mathbb{Q}(\delta)$. Performing the same steps as in the derivation of $J(x)$ we find that if we write $t_1, \ldots, t_6$ for the elementary symmetric polynomials in $\Psi_1, \ldots, \Psi_6$, this results in $t_1 = t_3 = 0$, $t_5 = -32f\delta$,

$$t_2 = 100T_2, \quad t_4 = 2000T_4, \quad t_6 = 40{,}000T_6$$

and

$$
\begin{aligned}
T_2 &= -3D^2 + 4CE, \\
T_4 &= 15D^4 - 40CD^2E + 20C^2E^2 - 4E^3 \\
&\quad + 8C^2Df + 6DEf - 2Cf^2, \\
T_6 &= -25D^6 + 100CD^4E - 100C^2D^2E^2 - 40C^2D^3f \\
&\quad + 80C^3DEf + 6D^3Ef - 12CDE^2f - 16C^4f^2 \\
&\quad - 2CD^2f^2 + 8C^2Ef^2 - E^2f^2 + Df^3.
\end{aligned} \tag{9.14}
$$

With these values, the sextic resolvent for Ψ is

$$x^6 + t_2x^4 + t_4x^2 + 32f\delta x + t_6,$$

with roots $\Psi_1, \ldots, \Psi_6$. For a solvable irreducible quintic we will have $\Psi = \rho_T\delta$ for some rational ρ_T and the sextic with root $\rho = \rho_T$ is

$$\Delta^3\rho^6 + t_2\Delta^2\rho^4 + t_4\Delta\rho^2 + 32\Delta f\rho + t_6, \tag{9.15}$$

with rational coefficients. Again, this will have a rational solution if and only if our quintic is solvable by radicals.

Note that

$$\oplus\left[\frac{1}{x_1}\left(\frac{1}{x_2} - \frac{1}{x_3}\right)\right] = \oplus\left[\frac{x_4x_5(x_3 - x_2)}{x_1x_2x_3x_4x_5}\right] = -\frac{1}{f}\oplus[x_1x_2(x_5 - x_4)] = -\frac{1}{f}\Psi,$$

where we have used the fact that $\oplus[x_1x_2x_3] = \oplus[x_1x_2x_5]$ in the final equality. Thus Ψ is $-f$ times the value of ψ that is obtained from the quintic whose roots are the reciprocals of $x_1, \ldots, x_5$, i.e. the quintic whose coefficients are reversed.

We can also consider the resolvent that is the ratio of ψ and Ψ, i.e.

$$\Lambda = \frac{\Psi}{\psi} = \frac{\oplus[x_1x_2(x_3 - x_4)]}{\oplus[x_1(x_2 - x_3)]}.$$

The resolvent Λ takes 6 values under the action of S_5 and is a root of the sextic polynomial

$$(\psi_1x - \Psi_1)(\psi_2x - \Psi_2)(\psi_3x - \Psi_3)(\psi_4x - \Psi_4)(\psi_5x - \Psi_5)(\psi_6x - \Psi_6), \tag{9.16}$$

which has rational coefficients. McClintock refers to the resolvent for ψ^2 as the *sinister* (i.e. left) resolvent, that for Ψ^2 as the *dexter* (i.e. right) resolvent and (9.16) as the central resolvent. The French mathematician George Perrin discovered this same resolvent independently a few years before McClintock. McClintock showed that the central resolvent can be written as

$$q(x)L(x) = 25Z(x)^2, \tag{9.17}$$

where q is the original quintic,

$$L(x) = L_0x + L_1 \quad \text{and} \quad Z(x) = G_0x^3 + G_1x^2 + G_2x + G_3. \tag{9.18}$$

The coefficients $L_0, L_1, G_0, G_1, G_2, G_3$ are give in terms of the coefficients of q via

$$\begin{gathered} G_0 = -C^3 + CE - D^2, \quad G_1 = -C^2D + Cf - DE, \\ G_2 = -CD^2 + C^2E + Df - E^2, \quad G_3 = 2CDE - C^2f - D^3 \end{gathered} \tag{9.19}$$

and

$$\begin{aligned} L_0 &= -15C^4E + 10C^3D^2 - 2C^2Df + 14C^2E^2 \\ &\quad - 22CD^2E + Cf^2 + 9D^4 - 2DEf + E^3, \\ L_1 &= 9C^4f - 20C^3DE + 10C^2D^3 + 8C^2Ef \\ &\quad - 12CDE^2 - 2CD^2f + 6D^3E + Df^2 - E^2f. \end{aligned}$$

Note, from (9.16), that the coefficient of x^6 in (9.17) must be a multiple of s_6 and the constant term must be a multiple of t_6. In fact

$$S_6 = L_0 - 25G_0^2, \quad T_6 = L_1f - 25G_3^2.$$

A slightly different type of resolvent is [19]

$$\theta = \oplus[x_1^2(x_2x_5 + x_3x_4)], \tag{9.20}$$

which takes only 6 values under the action of S_5. With θ_1 given by (9.20), the other 5 values can be written as

$$\begin{gathered} \theta_2 = (x_1x_2x_3)\theta_1, \quad \theta_3 = (x_1x_3x_2)\theta_1, \\ \theta_4 = (x_1x_2)\theta_1, \quad \theta_5 = (x_2x_3)\theta_1, \quad \theta_6 = (x_1x_3)\theta_1. \end{gathered}$$

Then $(\theta - \theta_1)(\theta - \theta_2)(\theta - \theta_3)(\theta - \theta_4)(\theta - \theta_5)(\theta - \theta_6)$ has coefficients which are all symmetric polynomials in $x_1, \ldots, x_5$ and hence in $\mathbb{Q}$. We will show below (in (9.31)) that

$$\theta = \tfrac{1}{4}(\psi^2 - c^2 - 12e). \tag{9.21}$$

9.3 McClintock's Method

We now turn to the actual solution of a quintic for which each of the sextic resolvents, equations (9.12) for ρ_S, (9.15) for ρ_T and (9.17) for Λ has a rational root. The method is sometimes described as Watson's method, though it was developed much earlier by McClintock. We assume that the irreducible quintic (9.1) is solvable by radicals and hence its Galois group is a subgroup of F_{20}. With the ordering of the roots implied by our particular choice of F_{20}, it is ψ_1^2 that is the rational root of (9.11). Of course, we can write the five roots of our quintic in any order we please, but only some of those orderings will result in ψ_1^2 being rational; others will result in ψ_2^2 or ψ_3^2, etc. being rational. There is no loss of generality in assuming $\psi_1^2 \in \mathbb{Q}$ and with this assumption, the order of the roots is then determined up to the action of $\sigma = (x_1x_2x_3x_4x_5)$ and $\tau = (x_2x_3x_5x_4)$. In other words, applying any permutation $\sigma^i\tau^j$ to the roots $x_1, \ldots, x_5$ will produce another valid order, but any other permutation in S_5 will not. We know that if $\delta = \sqrt{\Delta} \notin \mathbb{Q}$ then the Galois group of q is isomorphic to F_{20}. On the other hand, if $\delta \in \mathbb{Q}$, then either $G_q \cong D_{10}$ or $G_q \cong C_5$.

We write the roots of q in the form

$$\begin{aligned}
x_1 &= u_1 + u_2 + u_3 + u_4, \\
x_2 &= \zeta u_1 + \zeta^2 u_2 + \zeta^3 u_3 + \zeta^4 u_4, \\
x_3 &= \zeta^2 u_1 + \zeta^4 u_2 + \zeta u_3 + \zeta^3 u_4, \\
x_4 &= \zeta^3 u_1 + \zeta u_2 + \zeta^4 u_3 + \zeta^2 u_4, \\
x_5 &= \zeta^4 u_1 + \zeta^3 u_2 + \zeta^2 u_3 + \zeta u_4,
\end{aligned} \tag{9.22}$$

which is the equivalent for quintics of the expressions in (2.10) that we used for cubics. Since $1 + \zeta + \zeta^2 + \zeta^3 + \zeta^4 = 0$, this form ensures that the sum of the roots is zero and so there is no x^4 term in the quintic. The task of finding the five roots $x_1, \ldots, x_5$ is then reduced to that of determining the four quantities u_1, u_2, u_3, u_4. These can be expressed in terms of $x_1, \ldots, x_5$ via

$$\begin{aligned}
5u_1 &= x_1 + \zeta^4 x_2 + \zeta^3 x_3 + \zeta^2 x_4 + \zeta x_5, \\
5u_2 &= x_1 + \zeta^3 x_2 + \zeta x_3 + \zeta^4 x_4 + \zeta^2 x_5, \\
5u_3 &= x_1 + \zeta^2 x_2 + \zeta^4 x_3 + \zeta x_4 + \zeta^3 x_5, \\
5u_4 &= x_1 + \zeta x_2 + \zeta^2 x_3 + \zeta^3 x_4 + \zeta^4 x_5,
\end{aligned} \tag{9.23}$$

which can be compared with the cubic equivalents in (6.8). Note that

$$\sigma u_1 = \zeta u_1, \quad \sigma u_2 = \zeta^2 u_2, \quad \sigma u_3 = \zeta^3 u_3, \quad \sigma u_4 = \zeta^4 u_4 \tag{9.24}$$

and hence σ fixes u_1^5, u_2^5, u_3^5 and u_4^5. Also, we see that τ is equivalent to the cycle $(u_1u_3u_4u_2)$. The discriminant of $q(x)$ is $\Delta = \delta^2 \neq 0$ and we can take either square root for δ since $\tau\delta = -\delta$. This choice then fixes the order of the roots up to the action of σ and $\tau^2 = (x_2x_5)(x_3x_4) = (u_1u_4)(u_2u_3)$. For a solvable irreducible quintic it can be shown that $\Delta > 0$ so that δ is always real and we will take the positive square root, unless stated otherwise. In fact Δ is either a square in $\mathbb{Q}$ or a sum of two squares [19].

Expansion of $(x-x_1)(x-x_2)(x-x_3)(x-x_4)(x-x_5)$, with the roots $x_1, \ldots, x_5$ given by (9.22), and comparison with $x^5+10Cx^3+10Dx^2+5Ex+f$, shows that

$$-2C = u_2u_3 + u_1u_4, \tag{9.25}$$

$$-2D = u_1u_2^2 + u_1^2u_3 + u_4u_3^2 + u_2u_4^2, \tag{9.26}$$

$$-E = u_1^3u_2 - u_2^2u_3^2 + u_2^3u_4 - u_1^2u_4^2 + u_3u_4^3 + u_1u_3^3 + u_1u_2u_3u_4, \tag{9.27}$$

$$\begin{aligned}-f &= u_1^5 + u_2^5 + u_3^5 + u_4^5 \\ &\quad - 5(u_1u_4 - u_2u_3)(u_1^2u_3 - u_1u_2^2 - u_3^2u_4 + u_2u_4^2).\end{aligned} \tag{9.28}$$

Note that if we swap u_2 and u_3 and at the same time swap u_1 and u_4 (in other words apply τ^2), these equations remain unchanged. Thus, as we have already deduced from knowledge of the Galois group, the action of τ^2 leads to another valid ordering of the roots. Let

$$\begin{aligned} a_1 = u_1^2u_3, \quad a_2 = u_1u_2^2, \quad a_3 = u_3^2u_4, \quad a_4 = u_2u_4^2, \\ 2S = u_1u_4 - u_2u_3, \quad 2T = a_2 + a_3 - a_1 - a_4. \end{aligned} \tag{9.29}$$

The critical observation, on which the whole method rests, is that the quantities S and T are directly related to the resolvents ψ and Ψ. In fact we can substitute from (9.22) into the definitions of ψ and Ψ, (9.6) and (9.13), and compute

$$\psi = 5\sqrt{5}(u_1u_4 - u_2u_3) = 10\sqrt{5}S$$

and

$$\Psi = 5\sqrt{5}(u_1u_2^2 - u_1^2u_3 + u_4u_3^2 - u_2u_4^2) = 10\sqrt{5}T.$$

It is worth pointing out that none of the algebraic computations we have done to get this far has relied on our quintic being solvable. If it is solvable, then the sextic equations (9.12) and (9.15) have rational solutions and we then have

$$10\sqrt{5}S = \rho_S\delta, \quad 10\sqrt{5}T = \rho_T\delta \tag{9.30}$$

and S and T can therefore be considered known. However, if q is not solvable, then ρ_S and ρ_T cannot be determined algebraically and so in that case (9.30) is of no use.

We will assume that we have found the rational root ρ_S of (9.12) and the rational root ρ_T of (9.15). Later we will show that we can derive one from the other but for now we will consider them known quantities. Note that S^2 and T^2 are both rational, as is S/T provided $T \neq 0$. We can also compute θ given by (9.20) and find that

$$\theta = 5(3u_1^3u_2 + 3u_2^3u_4 + 3u_3u_4^3 + 3u_1u_3^3 + 2u_1^2u_4^2 + 2u_2^2u_3^2 - 12u_1u_2u_3u_4).$$

This can be reduced to

$$\theta = 125S^2 - 25C^2 - 15E, \tag{9.31}$$

from which (9.21) follows.

For notational convenience, we introduce the quantities

$$\alpha = S - C, \quad \beta = S + C, \quad \xi = T - D, \quad \eta = T + D$$

and we will initially assume that none of S, α or β is zero. From (9.25), (9.26) and (9.29), we have

$$u_1u_4 = \alpha, \quad u_2u_3 = -\beta, \quad a_1 + a_4 = -\eta, \quad a_2 + a_3 = \xi. \tag{9.32}$$

It then follows that

$$a_1a_4 = u_1^2u_3u_2u_4^2 = -\alpha^2\beta, \quad a_2a_3 = u_1u_2^2u_3^2u_4 = \alpha\beta^2, \tag{9.33}$$

and

$$a_1 - a_4 = R_1, \quad a_2 - a_3 = R_2, \tag{9.34}$$

where

$$R_1 = \sqrt{\eta^2 + 4\alpha^2\beta}, \quad R_2 = \sqrt{\xi^2 - 4\alpha\beta^2}. \tag{9.35}$$

We noted after (9.24) that σ fixes u_1^5, u_2^5, u_3^5, u_4^5. It also fixes a_1, a_2, a_3, a_4, S, T, α, β, ξ, η, R_1, R_2, θ and δ. On the other hand, τ is equivalent to the cycle $(u_1u_3u_4u_2)$ and also to $(a_1a_3a_4a_2)$. Hence

$$\begin{gathered}\tau S = -S, \quad \tau T = -T, \quad \tau\theta = \theta, \quad \tau\delta = -\delta, \\ \tau\alpha = -\beta, \quad \tau\beta = -\alpha, \quad \tau\xi = -\eta, \quad \tau\eta = -\xi, \\ \tau R_1 = -R_2, \quad \tau R_2 = R_1.\end{gathered}$$

The last pair of equations show us that the action of τ changes the sign of R_1R_2 and hence R_1R_2 is uniquely determined once we have fixed the sign of δ.

Ignoring the question of the appropriate sign of the square roots in (9.35) for the moment, once R_1 and R_2 have been determined, we have, from (9.32) and (9.34),

$$2a_{1,4} = -\eta \pm R_1, \quad 2a_{2,3} = \xi \pm R_2, \tag{9.36}$$

where the upper sign is taken to correspond to the first subscript in each case, a convention we adopt throughout. We can determine a_1, a_2, a_3 and a_4 from these equations and then we can compute u_1^5, u_2^5, u_3^5 and u_4^5, since, from (9.29) and (9.32),

$$\beta^2 u_1^5 = a_1^2 a_2, \quad \alpha^2 u_2^5 = a_2^2 a_4, \quad \alpha^2 u_3^5 = a_3^2 a_1, \quad \beta^2 u_4^5 = a_4^2 a_3. \tag{9.37}$$

These formulas give the fifth powers of the unknowns u_1, u_2, u_3, u_4 and so in order to obtain the roots of the quintic we need to extract fifth roots. However, just as in the choice of cube roots in the solution of the cubic, the choices of fifth roots are not independent. To remedy this, we note that we also have

$$\beta u_1^3 u_2 = -a_1 a_2, \quad \alpha u_2^3 u_4 = a_2 a_4, \quad \alpha u_1 u_3^3 = a_1 a_3, \quad \beta u_3 u_4^3 = -a_3 a_4. \tag{9.38}$$

Having chosen a fifth root for u_1, these equations allow us to compute u_2, u_3 and u_4 uniquely from

$$u_2 = \frac{a_4 u_1^2}{\alpha^2}, \quad u_3 = -\frac{a_3 a_4 u_1^3}{\alpha^3 \beta}, \quad u_4 = \frac{a_3 a_4^2 u_1^4}{\alpha^4 \beta^2}, \tag{9.39}$$

where we have also made use of (9.33). Since $u_1 u_2 u_3 u_4 = -\alpha\beta \neq 0$ by assumption, none of u_1, u_2, u_3 or u_4 can be zero. Hence (9.39) serves to determine u_2, u_3 and u_4 once u_1 is known. The five different choices of the fifth root for u_1 correspond to the action of the cyclic permutation σ on the order of the roots. Note that if u_1^5 is real, we can take u_1 to be real and then u_2, u_3 and u_4 will also be real, as will x_1.

The only thing that remains is to determine the signs of the square roots in the expressions for R_1 and R_2, for which we use (9.27), which is equivalent to

$$-E = u_1^3 u_2 + u_2^3 u_4 + u_3 u_4^3 + u_1 u_3^3 - 3S^2 - C^2. \tag{9.40}$$

If we multiply this equation by $\alpha\beta$, which is non-zero, and use (9.34) and (9.38), we can show that

$$S R_1 R_2 = \alpha\beta(E - 3S^2 - C^2) - C\xi\eta. \tag{9.41}$$

This shows that the signs of the square roots R_1 and R_2 must be chosen so that their product has the unique value given by (9.41). Changing the signs of both R_1 and R_2 is equivalent to swapping u_1 with u_4 and u_2 with u_3, i.e. applying τ^2. This also shows that if either R_1 or R_2 is zero, then we are free to choose either sign for the other. It is interesting to note that to get this far we have not made any use of (9.28), which can be rewritten in terms of S and T as

$$u_1^5 + u_2^5 + u_3^5 + u_4^5 = -f - 20ST. \tag{9.42}$$

In a sense we are done: if none of S, $S - C$ or $S + C$ is zero and if S and T have been calculated from their associated sextic resolvent equations, then the roots of the quintic are given by (9.22) with (9.37) used to determine u_1 and then u_2, u_3 and u_4 given by (9.39). However, any of the three quantities listed above may be zero and a solution method that doesn't cater for these cases is somewhat unsatisfactory. Moreover, as we shall see, S and T are not independent quantities and it will be useful to be able to express one in terms of the other, thus reducing the need to find a rational solution to a sextic to just one such equation.

We will now derive a pair of equations containing only S, T and the coefficients of the original quintic. The first is obtained by squaring (9.41) and replacing $R_1^2 R_2^2$ from (9.35). A factor of $\alpha\beta$ can be removed from the resulting equation to yield

$$\begin{aligned}\xi^2\eta^2 - 16DS^3T - 8CS^2(T^2 + D^2) + 2C\xi\eta(E - 3S^2 - C^2)&\\ = \alpha\beta\left[(C^2 - 5S^2)^2 + E^2 - 2E(3S^2 + C^2)\right].&\end{aligned} \tag{9.43}$$

This equation is of degree 6 in S and degree 4 in T.

The second equation is obtained from (9.42). If we write

$$\begin{aligned}X &= a_1^2 a_2 + a_4^2 a_3 = \beta^2(u_1^5 + u_4^5),\\ Y &= a_2^2 a_4 + a_3^2 a_1 = \alpha^2(u_2^5 + u_3^5),\end{aligned}$$

we have

$$\begin{aligned}\alpha^2\beta^2(f + 20ST) &= -\alpha^2 X - \beta^2 Y\\ &= -(S^2 + C^2)(X + Y) + 2SC(X - Y)\end{aligned} \tag{9.44}$$

and we can show from (9.36) that

$$\begin{aligned}X + Y &= D\xi\eta - TR_1R_2 + 2\alpha\beta(ST + CD),\\ X - Y &= T\xi\eta - DR_1R_2 - 2\alpha\beta(SD + TC).\end{aligned} \tag{9.45}$$

Hence (9.44) becomes

$$\alpha^2\beta^2(f+20ST) = -\xi\eta\left(D(S^2+C^2)-2CST\right) \\ + R_1R_2\left(T(S^2+C^2)-2SCD\right) \\ - 2\alpha\beta\left(S^3T+3S^2CD+3STC^2+C^3D\right). \tag{9.46}$$

If we now multiply by S and substitute for SR_1R_2 from (9.41) this reduces to

$$S\alpha\beta f = \xi\eta(CT-DS) + E(T(S^2+C^2)-2SCD) - T(C^2-5S^2)^2. \tag{9.47}$$

This equation is of degree 4 in S and degree 3 in T.

There are a number of ways of considering (9.43) and (9.47). If we find S and T from their respective resolvent equations, then they are simply identities that we can check. On the other hand, if we do not know S and T then they serve as a pair of simultaneous equations for S and T and the challenge is to find a pair of solutions such that S^2 and T^2 are rational. This is the way Young approached these equations in 1888 and he managed to solve some specific examples by inspection. Another possibility is that S is known from its resolvent equation and then we can use a combination of (9.43) and (9.47) to determine the appropriate value of T. This was how Watson tackled the problem in the 1940s and still involves a certain amount of trial and error though Watson was emphatic that "there is no difficulty at all in finding the single value of T that satisfies both [equations]" [4, p. 27]. A third approach, due to McClintock, and one which we will now elaborate on, is to use (9.43) and (9.47) to relate S and T directly and then once S is known, T follows. To this end we introduce the quantity

$$\Lambda = T/S,$$

which is the rational root of (9.17).

McClintock showed that if Λ is a root of the central resolvent, then

$$S^2 = -\frac{L(\Lambda)}{25Z(\Lambda)} = -\frac{Z(\Lambda)}{q(\Lambda)}, \tag{9.48}$$

where $L(x)$ and $Z(x)$ are defined in (9.18) and $q(x)$ is the quintic (9.1). We can also derive expressions for Λ in terms of S^2. To do this we write (9.43) and (9.47) as, respectively, a quartic and cubic in Λ with coefficients that are functions of S^2. We then need to determine their common rational root and we find that

$$\Lambda = \frac{\nu}{S_6 - S_4(5S)^2 - 3S_2(5S)^4 - (5S)^6}, \tag{9.49}$$

where

$$\begin{aligned}\nu = (2CD + f)(5S)^4 - (20C^3D + 22C^2f - 36CDE + 24D^3 + 2Ef)(5S)^2 \\ + 50C^5D - 59C^4f + 20C^3DE + 40C^2D^3 + 42C^2Ef \\ - 48CD^2f - 38CDE^2 + 44D^3E - Df^2 + E^2f\end{aligned}$$

and S_2, S_4, S_6 are as in (9.8).

At this point it will perhaps be helpful to summarise the process by which we can solve a solvable quintic when $S \neq 0, \pm C$.

- Either, compute S from its resolvent equation and then Λ is given by (9.49). Or, compute Λ from its resolvent equation and then S can be computed from (9.48). Then set $T = \Lambda S$.
- Compute R_1 from (9.35) and, provided $R_1 \neq 0$, R_2 from (9.41) so as to ensure the correct sign for R_1R_2. If $R_1 = 0$, then R_2 can be computed from (9.35), taking either sign for the square root.
- Next, from (9.37),

$$u_1^5 = \frac{(T + D - R_1)^2(T - D + R_2)}{8(S + C)^2}$$

  and we take any fifth root and compute u_2, u_3 and u_4 from (9.39).
- Finally, the roots are obtained from (9.22).

An alternative approach starts by defining

$$p_1 = u_1^5 + u_4^5, \quad p_2 = u_2^5 + u_3^5, \quad q_1 = u_1^5 - u_4^5, \quad q_2 = u_2^5 - u_3^5,$$
$$r_1 = p_1 + p_2, \quad r_2 = p_1 - p_2.$$

Equation (9.42) is $r_1 = -f - 20ST$ and we have

$$q_1 = \sqrt{p_1^2 - 4\alpha^5}, \quad q_2 = \sqrt{p_2^2 + 4\beta^5}. \tag{9.50}$$

Also

$$2u_1^5 = p_1 + q_1, \quad 2u_2^5 = p_2 + q_2, \quad 2u_3^5 = p_2 - q_2, \quad 2u_4^5 = p_1 - q_1, \tag{9.51}$$

showing that the choice of the square roots for q_1 and q_2 in (9.50) interchanges $u_1 \leftrightarrow u_4$ and $u_2 \leftrightarrow u_3$, respectively. Thus, in order to be able to determine an acceptable order for $u_1^5, u_2^5, u_3^5, u_4^5$, we also need to know the sign of q_1q_2. If we can find expressions for r_2 and q_1q_2, then we have all we need to compute u_1^5, u_2^5, u_3^5 and u_4^5.

Before deriving the necessary expressions, we will construct the quartic equation that has roots $u_1^5, u_2^5, u_3^5, u_4^5$, which we refer to as the quartic resolvent of the quintic:

$$(z - u_1^5)(z - u_2^5)(z - u_3^5)(z - u_4^5).$$

It was Leibniz who first conjectured that one could express the roots of a quintic as a sum of fifth roots of the solution of a resolvent quartic equation and Euler, Lagrange and Vandermonde all developed the idea further; see, for example, [3, §3]. By combining the first and fourth, and the second and third factors, we can write this as the product of the two quadratic factors

$$(z^2 - p_1 z + \alpha^5)(z^2 - p_2 z - \beta^5). \tag{9.52}$$

Once we have found r_2, we can easily determine p_1 and p_2 and then u_1^5 and u_4^5 are the roots of the first quadratic factor while u_2^5 and u_3^5 are the roots of the other factor. In order to obtain a valid order for the four roots, however, we still need to know the sign of q_1q_2 and refer back to (9.51).

We can calculate r_2 in much the same way that we derived (9.47). We have

$$\alpha^2\beta^2 r_2 = \alpha^2 X - \beta^2 Y = (S^2 + C^2)(X - Y) - 2SC(X + Y), \tag{9.53}$$

with $X = a_1^2 a_2 + a_4^2 a_3$ and $Y = a_2^2 a_4 + a_3^2 a_1$ as in (9.44). If we multiply by S and use (9.41) and (9.45), we can show that a factor of $\alpha\beta$ can be removed and this reduces to

$$\begin{aligned} S\alpha\beta r_2 &= \xi\eta(ST - CD) + E(2CST - D(S^2 + C^2)) \\ &\quad - 4CST(3S^2 + C^2) + D\alpha^2\beta^2. \end{aligned} \tag{9.54}$$

McClintock noted a remarkable simplification to this equation. If we take $S\times$(9.54) and add to it $C\times$(9.47), then it turns out that an additional factor of $\alpha\beta$ can be removed, leaving

$$S^2 r_2 = SG_1 + DS^3 + TG_0 + T^3 + 13CS^2T, \tag{9.55}$$

where G_0 and G_1 are defined in (9.19).

We can calculate q_1q_2 in terms of S and T using the same approach but with a slightly different form for X and Y. Thus

$$\begin{aligned} -\alpha\beta q_1 q_2 &= -\alpha\beta(u_1^5u_2^5 - u_1^5u_3^5 - u_2^5u_4^5 + u_3^5u_4^5) \\ &= S(\tilde{X} + \tilde{Y}) - C(\tilde{X} - \tilde{Y}), \end{aligned} \tag{9.56}$$

where $\tilde{X} = a_1 a_2^3 + a_4 a_3^3$ and $\tilde{Y} = a_1^3 a_3 + a_4^3 a_2$. Some lengthy algebra reveals that

$$\tilde{X} + \tilde{Y} = -\xi\eta(T^2 + D^2) - 2DTR_1R_2 + \alpha\beta\,(3C\xi\eta - SR_1R_2)\,,$$
$$\tilde{X} - \tilde{Y} = 2DT\xi\eta + R_1R_2(T^2 + D^2) + \alpha\beta\,(3S\xi\eta - CR_1R_2)\,.$$

If we use these expressions in (9.56), multiply by S and substitute for SR_1R_2 from (9.41), we find that

$$\begin{aligned} Sq_1q_2 &= \xi\eta\left(T^2 + D^2 - C\alpha\beta\right) \\ &\quad + (E - 3S^2 - C^2)\left(\alpha^2\beta^2 + C(T^2 + D^2) + 2DST\right). \end{aligned} \tag{9.57}$$

This is an equation which removes the sign ambiguity from the evaluation of q_2, once q_1 has been determined.

Before moving on to the special cases when $S = 0$ or $S = \pm C$, we give a number of examples. As far as possible, we have tried to use examples that have not been published before so as to increase the number of worked examples that are available. Other sources of examples include [4, 19, 40, 72, 74], though in many cases only a single root is given. In our examples, we will not prove at the outset that the quintics are irreducible and solvable, though the solutions that are derived demonstrate this. Most of the examples are generated using the methods described in Sect. 9.7 below.

Example 9.1

$$x^5 - 20x^3 + 170x + 208$$

Applying the test (9.5) shows that this quintic has one real root. The square root of the discriminant is $\delta = 14{,}400\sqrt{10}$ which is irrational and hence $G_q \cong F_{20}$. We find that

$$\begin{gathered} S = \sqrt{2}, \quad T = -4\sqrt{2}, \\ R_1 = 2\sqrt{4 - 2\sqrt{2}}, \quad R_2 = 2\sqrt{4 + 2\sqrt{2}}, \\ r_1 = -48, \quad r_2 = -48\sqrt{2}, \\ q_1 = 4\sqrt{50 + 31\sqrt{2}}, \quad q_2 = -4\sqrt{50 - 31\sqrt{2}}. \end{gathered}$$

The resolvent quartic is the product of the two factors

$$\left(z^2 + 24(1 \pm \sqrt{2})z + 232 \pm 164\sqrt{2}\right)$$

(continued)

Example 9.1 (continued)

and an appropriate ordering of the roots is

$$u_{1,4}^5 = 2\left(-6 - 6\sqrt{2} \pm \sqrt{50 + 31\sqrt{2}}\right),$$

$$u_{2,3}^5 = 2\left(-6 + 6\sqrt{2} \mp \sqrt{50 - 31\sqrt{2}}\right).$$

If we take u_1 to be one of the fifth roots of u_1^5 then, from (9.39), u_2, u_3 and u_4 can be determined via

$$u_2 = \left(-4 + 3\sqrt{2} - \sqrt{29 - \tfrac{41}{2}\sqrt{2}}\right) u_1^2,$$

$$u_3 = \left(-8 + \tfrac{11}{2}\sqrt{2} - \sqrt{116 - 82\sqrt{2}}\right) u_1^3,$$

$$u_4 = \left(21 - 15\sqrt{2} - \tfrac{1}{2}\sqrt{3554 - 2513\sqrt{2}}\right) u_1^4.$$

Finally, the roots of the quintic are then given by (9.22). All the square roots above are of positive quantities and hence if we are only interested in the real root of the quintic then we can simply take the real fifth root of u_1^5 and then all of u_1, u_2, u_3, u_4, and hence x_1, will be real. ♦

Example 9.2

$$x^5 - 20x^3 - 80x^2 - 165x - \tfrac{484}{3}$$

Applying the test (9.5) shows that there is one real root. The square root of the discriminant is $\delta = (2{,}807{,}200\sqrt{5})/9$ which is irrational and hence $G_q \cong F_{20}$. We find that

$$S = 1, \quad T = 2, \quad R_1 = 0, \quad R_2 = 2\sqrt{22},$$
$$r_1 = \tfrac{364}{3}, \quad r_2 = \tfrac{176}{3}, \quad q_1 = 18\sqrt{22}, \quad q_2 = \tfrac{20}{3}\sqrt{22}.$$

The quartic resolvent is

$$\tfrac{1}{3}(z^2 - 90z + 243)(3z^2 - 94z + 3).$$

(continued)

Example 9.2 (continued)

An appropriate ordering of the roots is

$$u_{1,4}^5 = 9(5 \pm \sqrt{22}), \quad u_{2,3}^5 = \tfrac{1}{3}(47 \pm 10\sqrt{22})$$

and we have

$$u_2 = \tfrac{1}{3}u_1^2, \quad u_3 = \tfrac{1}{9}(5 - \sqrt{22})u_1^3, \quad u_4 = \tfrac{1}{9}(5 - \sqrt{22})u_1^4.$$

If we set u_1 to be the real fifth root of $9\left(5 + \sqrt{22}\right)$, then the real root of the quintic is

$$x_1 = u_1 + \tfrac{1}{3}u_1^2 + \tfrac{1}{9}(5 - \sqrt{22})\left(u_1^3 + u_1^4\right). \qquad \blacklozenge$$

Example 9.3

$$x^5 + 40x^2 - 10x + 32$$

Applying the test (9.5) shows that there is one real root. The square root of the discriminant is $\delta = 177{,}600\sqrt{10}$ which is irrational and hence $G_q \cong F_{20}$. We find that

$$\begin{aligned} S &= \sqrt{2}, \quad T = 0, \\ R_1 &= 2\sqrt{4 + 2\sqrt{2}}, \quad R_2 = -2\sqrt{4 - 2\sqrt{2}}, \\ r_1 &= -32, \quad r_2 = 8\sqrt{2}, \\ q_1 &= 12\sqrt{2 - \sqrt{2}}, \quad q_2 = -12\sqrt{2 + \sqrt{2}}. \end{aligned}$$

The quartic resolvent is the product of the two factors

$$\left(z^2 + 4(4 \pm \sqrt{2})z \mp 4\sqrt{2}\right).$$

An appropriate ordering of the roots is

$$u_{1,4}^5 = -8 + 2\sqrt{2} \pm 6\sqrt{2 - \sqrt{2}}, \quad u_{2,3}^5 = -8 - 2\sqrt{2} \mp 6\sqrt{2 + \sqrt{2}}$$

(continued)

Example 9.3 (continued)

and we have

$$u_2 = \left(-1 - \tfrac{1}{2}\sqrt{4+2\sqrt{2}}\right)u_1^2,$$

$$u_3 = \left(-1 + \tfrac{1}{2}\sqrt{2} - \sqrt{2-\sqrt{2}}\right)u_1^3$$

$$u_4 = \left(-2 + \tfrac{1}{2}\sqrt{2} - \tfrac{3}{2}\sqrt{2-\sqrt{2}}\right)u_1^4.$$

♦

9.4 Special Cases

9.4.1 $S^2 = C^2 \neq 0$

If $S = \pm C$, then from (9.30),

$$\delta = \pm\frac{10C}{\rho s}\sqrt{5}, \tag{9.58}$$

which is irrational and so $G_q \cong F_{20}$. We also see that choosing a different sign for the square root of the discriminant results in a change in sign for S and so the two cases $S = C$ and $S = -C$ are essentially the same, related through the action of τ.

If either α or β is zero, some parts of (9.37) and (9.39) are no longer valid. It is helpful to look at the resolvent quartic (9.52) to see what is going on. If $S = C$ so that $\alpha = 0$, $\beta = 2C \neq 0$, the quartic is

$$z(z - p_1 z)(z^2 - p_2 z - 32C^5)$$

and we know from (9.32) that $u_1u_4 = 0$, $u_2u_3 = -2C$. It is thus apparent that at least one of u_1 and u_4 must be zero (and both will be zero if $p_1 = 0$). We are at liberty to choose which is always zero since the order of the roots is only determined up to an action of τ^2, so we will set $u_4 = 0$. It follows from (9.29) that a_3 and a_4 are zero. We then see, from (9.32) and (9.34), that $a_1 = -\eta = R_1$, $a_2 = \xi = R_2$ and the one part of (9.37) that we can still make use of is

$$4C^2u_1^5 = \xi\eta^2. \tag{9.59}$$

From (9.38) we have $2Cu_1^3u_2 = \xi\eta$ and hence, if we assume that $u_1 \neq 0$,

$$u_2 = \frac{2C}{\eta}u_1^2, \quad u_3 = -\frac{2C}{u_2} = -\frac{4C^2}{\xi\eta}u_1^3. \tag{9.60}$$

Example 9.4

$$x^5 + 10x^3 + 10x^2 + 40x - 54$$

Applying the test (9.5) shows that there is one real root. The square root of the discriminant is $\delta = 201{,}100\sqrt{5}$ and hence $G_q \cong F_{20}$. We have

$$S = 1 = C, \quad T = 3,$$
$$R_1 = 22\mathrm{i}\sqrt{50 + 10\sqrt{5}}, \quad R_2 = -22\mathrm{i}\sqrt{50 - 10\sqrt{5}},$$
$$r_1 = -6, \quad r_2 = 22, \quad q_1 = 8, \quad q_2 = 18,$$

and the resolvent quartic is

$$z(z-8)(z-2)(z+16),$$

with

$$u_1^5 = 8, \quad u_2^5 = 2, \quad u_3^5 = -16, \quad u_4 = 0.$$

If we set $u_2 = \sqrt[5]{2}$, the real root of the quintic is

$$x_1 = u_2 + u_2^3 - u_2^4. \qquad \blacklozenge$$

On the other hand, if $u_1 = u_4 = 0$, we have immediately from (9.25)–(9.28), that

$$D = 0, \quad E = u_2^2 u_3^2 = 4C^2, \quad f = -u_2^5 - u_3^5 = -p_2.$$

It follows from (9.14) that $T_6 = 0$ and hence that $T = 0$. The quantities u_2^5 and u_3^5 are the roots of the quadratic (taken in either order)

$$z^2 + fz - 32C^5. \tag{9.61}$$

This is an example of a De Moivre quintic; see Sect. 9.8.

The case $S = -C$ so that $\beta = 0$, $\alpha = -2C \neq 0$ is very similar. The quartic is

$$z(z - p_2)(z^2 - p_1 z - 32C^5)$$

and we know from (9.32) that $u_1 u_4 = -2C$, $u_2 u_3 = 0$. We choose to set $u_3 = 0$ and hence a_1 and a_3 are zero. We then have $a_2 = \xi = R_2$, $a_4 = -\eta = -R_1$ and, from (9.37) and (9.38), if we assume that $u_2 \neq 0$,

$$4C^2 u_2^5 = -\xi^2 \eta, \quad u_4 = -\frac{2C}{\xi} u_2^2, \quad u_1 = -\frac{2C}{u_4} = -\frac{4C^2}{\xi\eta} u_2^3.$$

If $u_2 = u_3 = 0$ we again have a De Moivre quintic. In this case,

$$D = 0, \quad E = u_1^2 u_4^2 = 4C^2, \quad f = -u_1^5 - u_4^5 = -p_1.$$

and thus the roots u_1^5 and u_4^5 are again the roots of the quadratic (9.61).

9.4.2 $S = 0, CT \neq 0$

If $S_6 = 0$, where S_6 is given in terms of the coefficients in the original quintic in (9.8), then $S = 0$. If we set $S = 0$ (which implies that $\alpha = -\beta = -C$) in (9.41) and (9.46), we obtain

$$\xi\eta + C(E - C^2) = 0$$

and

$$C^2 f + \xi\eta D - R_1 R_2 T - 2C^3 D = 0.$$

We deduce that in this case

$$T^2 = -G_0, \quad R_1 R_2 T = C G_1, \tag{9.62}$$

with G_0 and G_1 as defined in (9.19). Note that (9.10) shows that if $S = 0$ with non-zero C,

$$G_1^2 = T^2 \left((5C^2 - E)^2 + 16CD^2 \right).$$

When $S = 0$ we cannot evaluate Λ in order to determine T. However, from (9.62), there are only two possibilities and so we can just check which value of ρ_T (i.e. $\pm 10\sqrt{-5G_0/\delta}$) satisfies (9.15). We compute R_1 from (9.35) and, provided $T \neq 0$, R_2 from (9.62) so as to ensure the correct sign.

If we set $S = 0$ in (9.53) and use (9.62), we obtain

$$Tr_2 = 4CT^2 - C^4 + E^2 - Df. \tag{9.63}$$

This gives r_2 if $S = 0$ and $T \neq 0$. We can multiply (9.56) by T and substitute for $R_1 R_2 T$ from (9.62) to yield

$$-CTq_1q_2 = 2DT^2\xi\eta + CG_1\left(2T^2 + CE\right), \tag{9.64}$$

which removes the sign ambiguity from q_1q_2 when $S = 0$ and $CT \neq 0$.

The resolvent quartic (9.52) is

$$(z^2 - p_1 z - C^5)(z^2 - p_2 z - C^5),$$

where

$$2p_{1,2} = r_1 \pm r_2 = -f \pm \frac{1}{T}\left(4CT^2 - C^4 + E^2 - Df\right).$$

Since $C \neq 0$, none of $u_1^5, u_2^5, u_3^5, u_4^5$ is zero.

Example 9.5

$$x^5 - 5x^3 - 20x^2 - \tfrac{135}{4}x - 18$$

Applying the test (9.5) shows that there is one real root. The square root of the discriminant is $\delta = 600\sqrt{10}$ and hence $G_q \cong F_{20}$. We have

$$\begin{gathered} S = 0, \quad T = -\tfrac{1}{2}\sqrt{2}, \\ R_1 = \sqrt{4 + 2\sqrt{2}}, \quad R_2 = -\sqrt{4 - 2\sqrt{2}}, \\ r_1 = 18, \quad r_2 = -\tfrac{17}{2}\sqrt{2}, \\ q_1 = \tfrac{3}{2}\sqrt{52 - 34\sqrt{2}}, \quad q_2 = -\tfrac{3}{2}\sqrt{52 + 34\sqrt{2}} \end{gathered}$$

and the resolvent quartic is the product of the two factors

$$\left(z^2 - (9 \pm \tfrac{17}{4}\sqrt{2})z + \tfrac{1}{32}\right).$$

We have

$$\begin{aligned} u_{1,4}^5 &= \tfrac{1}{8}\left(36 - 17\sqrt{2} \pm 6\sqrt{52 - 34\sqrt{2}}\right), \\ u_{2,3}^5 &= \tfrac{1}{8}\left(36 + 17\sqrt{2} \mp 6\sqrt{52 + 34\sqrt{2}}\right) \end{aligned}$$

(continued)

Example 9.5 (continued)

and

$$u_2 = \left(4 + \sqrt{2} - 2\sqrt{4 + 2\sqrt{2}}\right) u_1^2,$$

$$u_3 = \left(14 - 8\sqrt{2} - 4\sqrt{20 - 14\sqrt{2}}\right) u_1^3,$$

$$u_4 = \left(72 - 34\sqrt{2} - 12\sqrt{52 - 34\sqrt{2}}\right) u_1^4. \quad \blacklozenge$$

Example 9.6

$$x^5 - 20x^3 + 180x + 64$$

Applying the test (9.5) shows that there is one real root. The square root of the discriminant is $\delta = 1{,}280{,}000\sqrt{5}$ which is irrational and hence $G_q \cong F_{20}$. We have

$$\begin{aligned} S = 0, \quad T = 8, \quad R_1 = R_2 = 4\sqrt{2}, \\ r_1 = -64, \quad r_2 = 96, \quad q_2 = 7q_1 = -56\sqrt{2}, \end{aligned}$$

and the resolvent quartic is

$$\left(z^2 - 16z + 32\right)\left(z^2 + 80z + 32\right).$$

We can compute

$$u_{1,4}^5 = 8 \mp 4\sqrt{2}, \quad u_{2,3}^5 = -40 \mp 28\sqrt{2}$$

and

$$u_2 = -\tfrac{1}{2}(2 + \sqrt{2})u_1^2, \quad u_3 = -\tfrac{1}{2}u_1^3, \quad u_4 = \tfrac{1}{4}(2 + \sqrt{2})u_1^4.$$

With $u_1 = \sqrt[5]{8 - 4\sqrt{2}}$, the real root of the quintic is

$$x_1 = u_1 - \tfrac{1}{2}u_1^3 + \tfrac{1}{4}(2 + \sqrt{2})(u_1^4 - 2u_1^2). \quad \blacklozenge$$

If $D = E = 0$, then since $S = 0$ we have, setting $S_6 = 0$ in (9.8),

$$25C^5 - f^2 = 0$$

and this shows that this can only happen if $C > 0$. If we write $C = g^2$, then the quintic is $x^5 + 10g^2x^3 + 5g^5$ and it is apparent that g is just a scale factor, since with $x = g\tilde{x}$ the quintic becomes $g^5(\tilde{x}^5 + 10\tilde{x}^3 + 5)$.

Example 9.7

$$x^5 + 10x^3 + 5.$$

This corresponds to taking $D = E = S = 0$ and $C = 1$. The discriminant is $\Delta = 5^7 59^2$ and so $G_q \cong F_{20}$. We know from (9.62) that $T^2 = C^3 = 1$, but if we go back to the resolvent sextic for T we can show that the correct solution is $T = -1$. From (9.35) and (9.62) we obtain

$$R_1^2 = R_2^2 = 5C^3 = 5 = -R_1 R_2.$$

Hence we take

$$R_1 = -R_2 = \sqrt{5}.$$

We also have $r_1 = -f = -5$ and, from (9.63), $r_2 = -3$ and so the quartic resolvent factorises as

$$\left(z^2 + 4z - 1\right)\left(z^2 + z - 1\right).$$

From (9.37) we find that a possible ordering of the roots is

$$u_{1,4}^5 = -(2 \pm \sqrt{5}), \quad u_{2,3}^5 = -\tfrac{1}{2}(1 \pm \sqrt{5}).$$

Thus

$$u_1 = -\sqrt[5]{2 + \sqrt{5}}$$

and from (9.39),

$$u_2 = \tfrac{1}{2}(1 - \sqrt{5})u_1^2, \quad u_3 = \tfrac{1}{2}(\sqrt{5} - 3)u_1^3, \quad u_4 = (\sqrt{5} - 2)u_1^4.$$

Since $c > 0$ in this case we know from (9.5) that there is one real root, and it is given by $x_1 = u_1 + u_2 + u_3 + u_4$. ♦

9.4.3 $S = T = 0, C \neq 0$

If both S and T are zero (with $C \neq 0$) we see, from (9.62), that $G_0 = G_1 = 0$. Hence, from (9.19),

$$E = C^2 + \frac{D^2}{C}, \quad f = CD + \frac{DE}{C} = \frac{D}{C^2}\left(2C^3 + D^2\right). \tag{9.65}$$

In this case, the discriminant is given by

$$\Delta = \frac{5^5}{C^8}\left(4C^3 + D^2\right)^6 \tag{9.66}$$

and so $G_q \cong F_{20}$. If we apply test (9.5) to quintics which satisfy (9.65) we find that $4C^3 + D^2 < 0$ is the condition for there being five real roots.

If $S = T = 0$ then the choice of square root of the discriminant plays no role in the solution, since changing the sign of δ no longer effects either S or T. In other words, the application of τ (i.e. the cycle $(u_1u_3u_4u_2)$) to a solution leads to another valid solution, which is not true in general. Also, in this case, both R_1 and R_2 are given by $\pm\sqrt{4C^3 + D^2}$ and we are at liberty to pick either sign. For definiteness, we will take $R_1 = R_2$. Then u_1, u_2, u_3, u_4 can be computed from (9.37) and (9.39) exactly as in the general case. We also have, from (9.53), $C^2r_2 = -DR_1R_2$ which leads to

$$r_2 = -\frac{D}{C^2}(4C^3 + D^2)$$

and, from (9.56) and (9.65),

$$q_1q_2 = -E(4C^3 + D^2).$$

Example 9.8 If we set $C = -2$, $D = 2$ in (9.65) we obtain a quintic with $S = T = 0$:

$$x^5 - 20x^3 + 20x^2 + 10x - 6.$$

Since $4C^3 + D^2 < 0$, there are five real roots. The square root of the discriminant is $\delta = 34{,}300\sqrt{5}$ and hence $G_q \cong F_{20}$. We can compute

$$R_1 = R_2 = 2\mathrm{i}\sqrt{7}, \quad r_1 = 6, \quad r_2 = 14, \quad q_2 = -2q_1 = 4\mathrm{i}\sqrt{7}$$

and the resolvent quartic is

$$(z^2 - 10z + 32)(z^2 + 4z + 32),$$

(continued)

Example 9.8 (continued)
with

$$u_{1,4}^5 = 5 \mp \mathrm{i}\sqrt{7}, \quad u_{2,3}^5 = -2 \pm 2\mathrm{i}\sqrt{7}.$$

When a quintic has five real roots, then any solution in radicals necessarily involves taking roots of complex quantities, just as in the case of three real roots to a cubic equation; the so called *Casus Irreducibilis*, see Sect. 4.3. ♦

9.4.4 $S = C = 0$

Note that when $S = C = 0$, we have, since $S_6 = 0$,

$$16D^4 - E^3 + 2DEf = 0. \tag{9.67}$$

One possibility is that $D = E = 0$, in which case the quintic is $x^5 + f = 0$ and then u_1 is any fifth root of $-f$ and $u_2 = u_3 = u_4 = 0$. In this case the discriminant is, from (9.4), $\Delta = 5^5 f^4$ and so δ is a rational multiple of $\sqrt{5}$. The only other possibility is that neither of D and E is zero, which we now assume. In this case there is a unique value of f given by (9.67) for which $S = C = 0$. If we use (9.67) to substitute for f in the expression (9.3) for the discriminant, with $C = 0$, we find that

$$\Delta = \frac{5^5(256D^8 - 176D^4E^3 - E^6)^2}{16D^4E^4}$$

and hence δ is again a rational multiple of $\sqrt{5}$. Hence, if $S = C = 0$, we know that $G_q \cong F_{20}$. Since $C = 0$, we know from (9.5) that any solvable quintic of this type will have one real root.

Since $\alpha = \beta = 0$ in this case, we see from (9.32) that one of each pair u_1, u_4 and u_2, u_3 must be zero. Hence we set $u_2 = u_4 = 0$ and if we insert these values into (9.26)–(9.28), we find that

$$u_1^5 + u_3^5 = -f, \quad u_1^5 u_3^5 = -4D^2E.$$

From (9.29), we then have $T = D$. Hence u_1^5 and u_3^5 are roots of the quadratic $x^2 + fx - 4D^2E$, namely

$$\tfrac{1}{2}(-f \pm \sqrt{f^2 + 16D^2E}).$$

Using (9.67), we have

$$f^2 + 16D^2E = \frac{(E^2 - Df)^2}{D^2}$$

and the two roots can then be shown to be

$$u_1^5 = \frac{8D^3}{E}, \quad u_3^5 = -\frac{E^2}{2D}.$$

We can take u_1 and u_3 to be the real fifth roots, since then $u_1^2 u_3$ is the real fifth root of $-32D^5$, i.e. $-2D$, as required by (9.26). We also have

$$r_2 = \frac{8D^3}{E} + \frac{E^2}{2D} = \frac{E^2}{D} - f.$$

Example 9.9

$$x^5 + x^2 + x + \tfrac{4}{25}$$

In this example, $D = 1/10$, $E = 1/5$ and so $u_1^5 = 1/25$ and $u_3^5 = -1/5$. The real root of the quintic is $u^2 - u$, where $u = 1/\sqrt[5]{5}$. ♦

Euler, in 1764, noted that the quintic

$$x^5 - 5abx^2 - 5a^2bx - a^3b - ab^2,$$

where neither a or b is zero, is solvable in radicals. The coefficients satisfy (9.67) and so this is an example in which $S = C = 0$. In fact this is just another way of expressing the general case, with $a = E/(2D)$ and $b = -4D^2/E$. We have

$$u_1^5 = ab^2, \quad u_3^5 = a^3b$$

and the real root of the quintic is

$$x_1 = \sqrt[5]{ab^2} + \sqrt[5]{a^3b}.$$

Example 9.10 With $a = 1$ and $b = -2$, we find that the real root of

$$x^5 + 10x^2 + 10x - 2$$

is $\sqrt[5]{4} - \sqrt[5]{2}$. With $a = b = 1$, we see that 2 is a root and hence obtain the factorisation

$$x^5 - 5x^2 - 5x - 2 = (x - 2)(x^4 + 2x^3 + 4x^2 + 3x + 1).$$ ♦

9.5 D_{10} or C_5?

If $\delta \in \mathbb{Q}$, which we will assume in this section, then the Galois group of a solvable quintic is either D_{10} or C_5. If our quintic has a single real root, then we know from Proposition 7.4 that its Galois group wll be D_{10}. We will now give a number of different ways that can be used to make the determination of which of D_{10} or C_5 is the correct Galois group for quintics with five real roots. First, we know from (9.30) that in this case S and T are each rational multiples of $\sqrt{5}$ and $r_1 = -f - 20ST$ is rational. We also note, from Sects. 9.4.3 and 9.4.4, that if $S = T = 0$ or $S = C = 0$, then the discriminant is not a square in $\mathbb{Q}$ and so neither of these cases are possible. Thus either we have $S \neq 0$, in which case (9.55) shows that r_2 is a rational multiple of $\sqrt{5}$, or we have $S = 0$ with $CT \neq 0$, in which case the same result follows from (9.63).

The first approach (from [19]) is based on a generalisation of an approach we took to the cubic, described in Sect. 6.5. From (9.23), using the fact that $\zeta^5 = 1$, we can derive

$$\begin{aligned}
(5u_1)^5 &= \ell_0 + \ell_1\zeta + \ell_2\zeta^2 + \ell_3\zeta^3 + \ell_4\zeta^4,\\
(5u_2)^5 &= \ell_0 + \ell_1\zeta^2 + \ell_2\zeta^4 + \ell_3\zeta + \ell_4\zeta^3,\\
(5u_3)^5 &= \ell_0 + \ell_1\zeta^3 + \ell_2\zeta + \ell_3\zeta^4 + \ell_4\zeta^2,\\
(5u_4)^5 &= \ell_0 + \ell_1\zeta^4 + \ell_2\zeta^3 + \ell_3\zeta^2 + \ell_4\zeta,
\end{aligned}$$

where $\ell_0, \ldots, \ell_4$ are complicated expressions each of which is a sum of terms containing products of five of the roots $x_1, \ldots, x_5$. They are given explicitly in equations (5.0)–(5.4) of [19].

Each of the quantities $\ell_1, \ell_2, \ell_3, \ell_4$ is fixed by σ, but τ is equivalent to the cycle $(\ell_1\ell_2\ell_4\ell_3)$. Also, we can show that

$$\ell_0 + \ell_1 + \ell_2 + \ell_3 + \ell_4 = (x_1 + x_2 + x_3 + x_4 + x_5)^5,$$

which is zero for the depressed quintic. If we substitute for the roots in terms of $u_1, \ldots, u_4$, using (9.22), we can show that $\ell_0 = 625r_1$, which is rational, and

$$\ell_{1,4} = \tfrac{625}{4}[\sqrt{5}\,r_2 - r_1 \mp 2(\zeta - \zeta^4)\,(q_1 + \eta_0 q_2)],$$
$$\ell_{2,3} = \tfrac{625}{4}[-\sqrt{5}\,r_2 - r_1 \mp 2(\zeta - \zeta^4)\,(\eta_0 q_1 - q_2)],$$

where $\eta_0 = \left(-1 + \sqrt{5}\right)/2$. Note that $2(\zeta - \zeta^4) = \mathrm{i}\sqrt{10 + 2\sqrt{5}}$.

We now use the following result [19, Thm. 2]. If $\delta \in \mathbb{Q}$ then $G_q \cong C_5$ if and only if $\ell_1, \ell_2, \ell_3, \ell_4$ are rational, with $G_q \cong D_{10}$ otherwise. The determination of the Galois group thus boils down to proving the rationality or otherwise of the quantities

$$\sigma_1 = \mathrm{i}\sqrt{10 + 2\sqrt{5}}\left(2q_1 + (-1 + \sqrt{5})q_2\right),$$
$$\sigma_2 = \mathrm{i}\sqrt{10 + 2\sqrt{5}}\left((-1 + \sqrt{5})q_1 - 2q_2\right).$$

Since

$$q_1^2 = \tfrac{1}{4}(r_1 + r_2)^2 - 4(S - C)^5, \quad q_2^2 = \tfrac{1}{4}(r_1 - r_2)^2 + 4(S + C)^5,$$

with r_1, C rational and r_2, S rational multiples of $\sqrt{5}$, and noting (9.57) and (9.64) which show that $q_1 q_2$ is a rational multiple of $\sqrt{5}$, we see that

$$q_1^2 = \rho_1 + \rho_2\sqrt{5}, \quad q_2^2 = \rho_1 - \rho_2\sqrt{5}, \quad q_1 q_2 = \rho_3\sqrt{5}, \tag{9.68}$$

for some rational numbers ρ_1, ρ_2 and ρ_3. A straightforward calculation shows that we then have

$$\sigma_1^2 = -80(\rho_1 + \rho_2 + 2\rho_3), \quad \sigma_2^2 = -80(\rho_1 - \rho_2 - 2\rho_3).$$

Also, since

$$q_1^2 q_2^2 = \rho_1^2 - 5\rho_2^2 = 5\rho_3^2,$$

we find that

$$\sigma_1^2\sigma_2^2 = 80^2\left(5\rho_2^2 + 5\rho_3^2 - (\rho_2 + 2\rho_3)^2\right) = 80^2(2\rho_2 - \rho_3)^2$$

and so $\sigma_1\sigma_2 \in \mathbb{Q}$. Thus $\sigma_1 \in \mathbb{Q}$ implies $\sigma_2 \in \mathbb{Q}$ and vice versa. Hence the condition that $\ell_1, \ell_2, \ell_3, \ell_4$ are all rational is that

$$-5(\rho_1 + \rho_2 + 2\rho_3) \tag{9.69}$$

is a square in $\mathbb{Q}$.

Another test to determine whether G_q is isomorphic to D_{10} or C_5 when $\delta \in \mathbb{Q}$ is as follows. Let

$$Z_1 = (x_1 - x_2)(x_2 - x_3)(x_3 - x_4)(x_4 - x_5)(x_5 - x_1),$$

which is fixed by $\sigma = (x_1x_2x_3x_4x_5)$. With $\tau = (x_2x_3x_5x_4)$ we have

$$Z_2 = \tau Z_1 = (x_1 - x_3)(x_3 - x_5)(x_5 - x_2)(x_2 - x_4)(x_4 - x_1),$$

again fixed by σ, and $\tau Z_2 = \tau^2 Z_1 = -Z_1$. The quantity

$$X = Z_1^2 + Z_2^2$$

is fixed by both σ and τ and hence by F_{20} and it can be shown that X is not fixed by any other elements of S_5. By exactly the same argument that we used after (9.11), it follows that for any solvable irreducible quintic, X must be the rational root of a sextic resolvent with rational coefficients (which can be computed, though it is hideously complicated) and we have, from the definition of the discriminant,

$$\Delta = Z_1^2 Z_2^2.$$

Thus the sum and product of Z_1^2 and Z_2^2 are rational and they are the two roots of the quadratic (for Z^2)

$$\Delta = Z^2(X - Z^2), \tag{9.70}$$

with rational coefficients.

The subgroup of S_5 which fixes Z_1^2 is $D_{10} = \langle \sigma, \tau^2 \rangle$ and so the roots of (9.70) for Z^2 are rational if and only if the Galois group is contained within D_{10}. As we have seen, this is always the case for a solvable, irreducible quintic with $\delta \in \mathbb{Q}$. Similarly, the subgroup of S_5 which fixes Z_1 is $C_5 = \langle \sigma \rangle$ and $G_q \cong C_5$ if and only if the roots of (9.70) for Z^2 are squares in $\mathbb{Q}$. These observations are noted in [5]. (The product of the roots is $\Delta = \delta^2$, so if one is a square in $\mathbb{Q}$, so is the other.) Thus to test whether the Galois group is C_5 we can find X as the rational root of a sextic and then solve the quadratic (9.70). If this has roots which are squares of rational numbers, then $G_q \cong C_5$.

However, if we have solved for the roots, then we can compute Z directly and check whether or not it is rational. Substituting from (9.22), we can show that

$$\tfrac{1}{5} Z_1 = (\zeta^2 - \zeta^3)(\gamma R_1 + 2q_1 - q_2) - (\zeta - \zeta^4)(\gamma R_2 + q_1 + 2q_2) \tag{9.71}$$

where $\gamma = (c + \rho_S \delta)/2$. When calculating Z_1, it is important to ensure that the signs of R_1, R_2 and q_1, q_2 correspond to the same ordering of the roots.

Yet another test to determine when a quintic has Galois group C_5 is as follows [10, §6.3.4]. If we let

$$W = \oplus[x_1x_2(x_1 - x_2)]$$

then as long as $W \neq 0$, the Galois group is C_5 if and only if W^2 is rational. If $W = 0$ then extra work is required to transform the quintic into another with the same Galois group but for which $W \neq 0$. From (9.22) we can show that

$$W = \tfrac{5}{2}(\zeta - \zeta^4)\left((5 - \sqrt{5})R_1 + 2\sqrt{5}R_2\right). \tag{9.72}$$

9.6 Bring-Jerrard Quintics

In Chap. 8 we described how Bring and Jerrard showed that the problem of finding the roots of any quintic can be reduced, using a process that requires nothing worse than solving a cubic equation, to that of solving a quintic that has no x^4, x^3 or x^2 terms. We can solve such quintics using the results derived above, setting $C = D = 0$. In addition, as shown by Young [73], it is possible to parametrise all solvable, irreducible Bring-Jerrard quintics in terms of an arbitrary pair of rational numbers.

Assume that $C = D = 0$ and consider the quintic

$$q(x) = x^5 + 5Ex + f, \quad E \neq 0,\ f \neq 0,$$

for which $\Delta = 5^5(256E^5 + f^4)$. This shows that there will be multiple roots if and only if $E^5 + (f/4)^4 = 0$. Then, with $E = -a^4$ and $f = 4a^5$, we have

$$x^5 - 5a^4x + 4a^5 = (x - a)^2(x^3 + 2ax^2 + 3a^2x + 4a^3). \tag{9.73}$$

It is also possible that the quintic factorises into an irreducible quadratic and an irreducible cubic, an example being

$$x^5 + x + 1 = (x^2 + x + 1)(x^3 - x^2 + 1).$$

A complete list of possible factorisations (over $\mathbb{Q}$) of quintics of the form $x^5 \pm x + n$, where n is an integer, is given in [56].

We will assume that our quintic is irreducible and solvable by radicals. Since C is not negative, all solvable, irreducible Bring-Jerrard quintics have a single real root. In this case, the resolvent equation (9.11) can be shown to reduce to

$$(25S^2 - E)^4(25S^4 - 6ES^2 + E^2) = f^4S^2. \tag{9.74}$$

Since $E \neq 0$, we cannot have $S = 0$ and S^2 is rational since our quintic is solvable. We have $\alpha = \beta = S$ and $\xi = \eta = T$. Equations (9.43) and (9.47) become, respectively,

$$\begin{aligned} T^4 - 16S^6 &= S^2(3S^2 - E)^2, \\ Sf &= T(E - 25S^2). \end{aligned} \tag{9.75}$$

The first of these is simply the square of (9.41) and the two combined are equivalent to (9.74). Since neither S nor f is zero, we have $T \neq 0$, $25S^2 \neq E$, and

$$T = \frac{Sf}{E - 25S^2}.$$

Define rational numbers

$$\Omega = \frac{E - 3S^2}{4S^2}, \qquad \Lambda = \frac{f}{E - 25S^2} = \frac{T}{S}. \tag{9.76}$$

We then have, using (9.75),

$$\Lambda^4 = \frac{T^4}{S^4} = \frac{(3S^2 - E)^2 + 16S^4}{S^2} = 16S^2w^2,$$

where we have written

$$w = \sqrt{\Omega^2 + 1},$$

assumed positive, and from (9.76),

$$S^2 = \frac{E}{3 + 4\Omega} = \frac{f}{\Lambda(4\Omega - 22)}.$$

Hence we obtain

$$E = \frac{\Lambda^4}{16w^2}(3 + 4\Omega), \qquad f = \frac{\Lambda^5}{8w^2}(2\Omega - 11). \tag{9.77}$$

This is equivalent to the parametrisation derived by Young [73] (see also [26] and the discussion in [48, §§30–32]) and to those given in [65] and, less trivially, [13, Thm. 13.2.12]. Young showed that the quintic is solvable if and only if we can find rational quantities Ω and Λ such that E and f are given by (9.77). Note that Λ can be removed by a simple scaling: if we write $x = \Lambda\tilde{x}$, then our quintic becomes

$$\Lambda^5\left(\tilde{x}^5 + \frac{5(3 + 4\Omega)}{16w^2}\tilde{x} + \frac{2\Omega - 11}{8w^2}\right).$$

Hence we could fix Λ, though we would then need to remember that for each Ω, there is an infinite family of solvable quintics. With E and f given by (9.77), the discriminant of $x^5 + 5Ex + f$ is

$$\Delta = 5^5 \Lambda^{20} \frac{(4\Omega^3 + 84\Omega^2 - 37\Omega + 122)^2}{2^{12} w^{10}}, \tag{9.78}$$

which shows that in this case $\delta \in \mathbb{Q}\left(\sqrt{5}, w\right)$.

Since all solvable Bring-Jerrard quintics have a single real root, the Galois group cannot be C_5 (see Proposition 7.4). Hence the Galois group of a solvable, irreducible Bring-Jerrard quintic is either F_{20}, if the discriminant is not a square in $\mathbb{Q}$, or the dihedral group D_{10}, if the discriminant is a square in $\mathbb{Q}$. Looking at (9.78) we see that this is equivalent to

$$G_q \cong \begin{cases} D_{10}, & 5w^2 \text{ is a square in } \mathbb{Q}, \\ F_{20}, & 5w^2 \text{ is not a square in } \mathbb{Q}. \end{cases}$$

In terms of Ω and Λ we have $S = \Lambda^2/4w$, $T = \Lambda^3/4w$,

$$R_1^2 = \frac{\Lambda^6}{16w^2}\left(1 + \frac{1}{w}\right), \quad R_2^2 = \frac{\Lambda^6}{16w^2}\left(1 - \frac{1}{w}\right), \quad R_1 R_2 = \frac{\Lambda^6 \Omega}{16w^3}$$

and we can take $R_1 > 0$ and then R_2 has the same sign as Ω. We can also show that

$$r_1 = -\frac{\Lambda^5(2\Omega - 1)}{8w^2}, \qquad r_2 = \frac{\Lambda^5}{4w},$$

$$p_{1,2} = \frac{\Lambda^5}{16w^2}(1 - 2\Omega \pm 2w), \quad q_{1,2}^2 = \frac{\Lambda^{10}}{256w^4}\left[(1 - 2\Omega \pm 2w)^2 \mp \frac{1}{w}\right]$$

and $q_1 q_2 > 0$. We can evaluate a_1, a_2, a_3, a_4 and then

$$u_1^5 = \frac{\Lambda^5}{32w^2}(1 - 2\Omega + 2w - V),$$

where

$$V = 2w\sqrt{1 + \frac{1}{w}} - \operatorname{sgn}(\Omega)(2w + 1)\sqrt{1 - \frac{1}{w}}. \tag{9.79}$$

Note that when $\Omega = 0$, $w = 1$ and so there is no need to define $\operatorname{sgn}(0)$. It is not difficult to show that V is always positive. If we square (9.79), we find that

$$V^2 = 8\Omega^2 - 4\Omega + 5 + \frac{1}{w}\left(-8\Omega^3 + 4\Omega^2 - 8\Omega + 3\right).$$

Note that the square bracket in the expression for q_1^2 can be shown to be V^2.

From (9.39) we have

$$u_2 = -\frac{2}{\Lambda}(w + \sqrt{w^2 + w})u_1^2,$$

$$u_3 = \frac{4w}{\Lambda^2}(w - \Omega + \sqrt{2w(w - \Omega)})u_1^3,$$

$$u_4 = \frac{8w^2}{\Lambda^3}(1 - 2\Omega + 2w + V)\,u_1^4.$$

These, when substituted into (9.22), are explicit expressions for the roots of any solvable Bring-Jerrard quintic in terms of the parameters Ω and Λ. For example, if we write u for the real fifth root of $(1 - 2\Omega + 2w - V)/w^2$, the real root of the quintic is

$$x_1 = \frac{\Lambda}{2}\Big(u - u^2\left(w + \sqrt{w^2 + w}\right) + u^3 w\left(w - \Omega + \sqrt{2w(w - \Omega)}\right)$$
$$+ u^4 w^2 (1 - 2\Omega + 2w + V)\Big).$$

The resolvent quartic can be written, from (9.52), as the product of the two factors

$$\left(y^2 - (1 - 2\Omega \pm 2w)\,y \pm \frac{1}{4w}\right)$$

where, for convenience, we have set $z = \Lambda^5 y/(16w^2)$ so that the roots are $16w^2(u_1/\Lambda)^5$, etc.

Example 9.11 There will be significant simplification if we choose Ω such that $w^2 = \Omega^2 + 1$ is a square. This is equivalent to setting $\Omega = p/q$, where $p^2 + q^2 = r^2$ is a Pythagorean triple of integers. For example, we can take $\Omega = 15/8$, so that $w = 17/8$. This then leads to

$$V = \tfrac{11}{34}\sqrt{17}.$$

If we choose $\Lambda = 17$, then the coefficients in the quintic are integers:

$$x^5 + 60{,}690x - 284{,}954.$$

(continued)

Example 9.11 (continued)

This is a solvable Bring-Jerrard quintic and so it has one real root. The square root of the discriminant is $\delta = 6{,}802{,}676{,}872{,}700\sqrt{5}$ and so $G_q \cong F_{20}$. We find

$$\begin{gathered} S = 34, \quad T = 578, \\ R_1 = 170\sqrt{17}, \quad R_2 = 102\sqrt{17}, \\ r_1 = -108{,}086, \quad r_2 = 167{,}042, \\ q_1 = -6358\sqrt{17}, \quad q_2 = -33{,}524\sqrt{17} \end{gathered}$$

and the resolvent quartic factorises as

$$(z^2 - 29{,}478z + 45{,}435{,}424)(z^2 + 137{,}564z - 45{,}435{,}424).$$

Then

$$u_{1,4}^5 = 289(51 \mp 11\sqrt{17}), \quad u_{2,3}^5 = 578(-119 \mp 29\sqrt{17})$$

and

$$\begin{gathered} u_2 = -\tfrac{1}{4}(1 + \tfrac{5}{17}\sqrt{17})u_1^2, \quad u_3 = \tfrac{1}{136}(1 + \sqrt{17})u_1^3, \\ u_4 = \tfrac{1}{4624}(51 + 11\sqrt{17})u_1^4. \end{gathered}$$

♦

Example 9.12 If we take $\Omega = -7/24$, we have $w = 25/24$. This then leads to

$$V = \tfrac{53}{15}.$$

If we choose $\Lambda = 5$, then the coefficients in the quintic are integers:

$$x^5 + 330x - 4170.$$

This is a solvable Bring-Jerrard quintic and so it has one real root. The square root of the discriminant is $\delta = 434{,}952{,}900\sqrt{5}$ and so $G_q \cong F_{20}$. We have

$$\begin{gathered} S = 6, \quad T = 30, \quad R_1 = 42, \quad R_2 = -6, \\ r_1 = 570, \quad r_2 = 750, \quad q_1 = -636, \quad q_2 = -198 \end{gathered}$$

(continued)

Example 9.12 (continued)
and the resolvent quartic factorises as

$$(z-648)(z-54)(z-12)(z+144).$$

We can calculate

$$u_1^5 = 12, \quad u_2^5 = -144, \quad u_3^5 = 54, \quad u_4^5 = 648.$$

Equation (9.39) yields $u_2 = -u_1^2$, $u_3 = u_1^3/2$, $u_4 = u_1^4/2$, though in this case these formulas can be deduced by inspection. The real root of the quintic is given by

$$x_1 = u_1 - u_1^2 + \tfrac{1}{2}\left(u_1^3 + u_1^4\right), \quad u_1 = \sqrt[5]{12}.$$ ♦

An alternative parametrisation (given in [58]), which always generates Bring-Jerrard quintics with Galois group D_{10} can be obtained as follows. First we write

$$\Omega = \frac{u-4}{2(u+1)}$$

and note immediately that any rational Ω can be written in this way with $u \in \mathbb{Q}$, except $\Omega = 1/2$. Quintics with $\Omega = 1/2$ are

$$x^5 + \tfrac{5}{4}\Lambda^4 x - \Lambda^5 \tag{9.80}$$

and have Galois group D_{10} since $5w^2 = 25/4 = (5/2)^2$. The case $\Lambda = 1$ is:

Example 9.13

$$x^5 + \tfrac{5}{4}x - 1$$

This is a solvable Bring-Jerrard quintic and so it has one real root. The square root of the discriminant is $\delta = 125/2$. We find

$$\begin{gathered}
S = T = \tfrac{1}{10}\sqrt{5}, \\
R_1 = \tfrac{1}{10}\sqrt{5+2\sqrt{5}}, \quad R_2 = \tfrac{1}{10}\sqrt{5-2\sqrt{5}}, \\
r_1 = 0, \quad r_2 = \tfrac{1}{10}\sqrt{5}, \\
q_1 = -\tfrac{1}{100}\sqrt{125-10\sqrt{5}}, \quad q_2 = -\tfrac{1}{100}\sqrt{125+10\sqrt{5}}
\end{gathered}$$

(continued)

Example 9.13 (continued)

and the resolvent quartic (scaled so the roots are $200u_1^5$, etc.) is

$$(z^2 + 10\sqrt{5}z - 10\sqrt{5})(z^2 - 10\sqrt{5}z + 10\sqrt{5}).$$

Then

$$u_{1,4}^5 = \frac{5 \mp \sqrt{25 - 2\sqrt{5}}}{40\sqrt{5}}, \quad u_{2,3}^5 = \frac{-5 \mp \sqrt{25 + 2\sqrt{5}}}{40\sqrt{5}}$$

and

$$\begin{aligned} u_2 &= \left(-\sqrt{5} - \sqrt{5 + 2\sqrt{5}}\right) u_1^2, \\ u_3 &= \left(5 - \sqrt{5} + \sqrt{50 - 10\sqrt{5}}\right) u_1^3, \\ u_4 &= 2\sqrt{5}\left(5 + \sqrt{25 - 2\sqrt{5}}\right) u_1^4. \end{aligned}$$

♦

For $\Omega \neq 1/2$, we obtain

$$E = \frac{\Lambda^4(u^2 - 1)}{4(u^2 + 4)}, \quad f = -\frac{\Lambda^5(u + 1)(2u + 3)}{2(u^2 + 4)}$$

with

$$\Delta = \frac{5^6 \Lambda^{20}(u + 1)^4(2u^3 + 4u^2 + 11u + 8)^2}{16(u^2 + 4)^5}.$$

The condition for $G_q \cong D_{10}$ is now that $u^2 + 4$ is a square in $\mathbb{Q}$ and this is true if and only if

$$u = v - 1/v$$

so that $u^2 + 4 = (v + 1/v)^2$. We then have

$$\begin{aligned} E &= \frac{\Lambda^4(v^2 + v - 1)(v^2 - v - 1)}{4(v^2 + 1)^2}, \\ f &= -\frac{\Lambda^5(v^2 + v - 1)(v + 2)(2v - 1)}{2(v^2 + 1)^2}. \end{aligned}$$

Setting $v = 0$ generates (9.80) and so all Bring-Jerrard quintics with Galois group D_{10} are included in this parametrisation.

9.7 Generating Solvable Quintics

Most quintics are not solvable by radicals. Whilst this statement is imprecise, its meaning is fairly clear. If we select a quintic with rational coefficients "at random", then the chances are that it will not be solvable. To give a sense of what this means quantitatively, all quintics of the form $\alpha_5 x^5 + \alpha_3 x^3 + \alpha_2 x^2 + \alpha_1 x + \alpha_0$, for integers $\alpha_0, \alpha_5 = 1, \ldots, N$ and $\alpha_1, \alpha_2, \alpha_3 = -N, \ldots, N$ were tested for irreducibility and solvability. Allowing α_0 and α_5 to be negative gives nothing new since if they are both negative we will generate duplicates of those in the above list and with one negative, the change from x to $-x$ will lead to an equation in the list. For $N = 9$ we find that of the 555,579 quintics tested, the vast majority (512,879) are irreducible. Of the irreducible quintics, only 229 are solvable and of those, 72 are of the form $x^5 + f$. It is thus highly desirable to have a method for generating solvable quintics, since they are few and far between.

McClintock provided one such method. Choose four rational numbers Θ, Γ, Λ and Ω, write $w^2 = \Omega^2 + 1$ as in the previous section, and set

$$\chi = \frac{\Lambda^2\Omega^2 - \Gamma^2 w^2}{4(\Theta + 1)(\Theta^2 w^2 - 1)}, \tag{9.81}$$

which is also rational. Clearly we must avoid $\Theta = -1$ and $\Theta^2 = 1/w^2$ and we will also insist that $\chi \neq 0$. Now let

$$S^2 = \chi^2/w^2, \quad C = \Theta\chi, \quad D = (\Gamma + \Lambda)\chi \tag{9.82}$$

and then calculate E from

$$E = C^2 + S^2(3 + 4\Omega) + \frac{C(\Lambda^2 S^2 - D^2) + 2D\Omega\Lambda S^2}{S^2 - C^2}. \tag{9.83}$$

Finally, calculate f from (9.47), which if we identify Λ with T/S is

$$\begin{aligned}(S^2 - C^2)f &= (\Lambda^2 S^2 - D^2)(C\Lambda - D) \\ &\quad + E(\Lambda(S^2 + C^2) - 2CD) - \Lambda(C^2 - 5S^2)^2.\end{aligned}$$

We claim that $x^5 + 10Cx^3 + 10Dx^2 + 5Ex + f$ is a solvable quintic with $S \neq 0$, $S^2 \neq C^2$. Since, by construction, S^2 and Λ are rational, we just need to show that (9.81), (9.82) and (9.83) are consistent with the equations that we derived in Sect. 9.3.

With $\Lambda = T/S$, the relationship between Θ, Γ, Ω and the coefficients of the quintic is as follows. Let $\mu = R_1/R_2$ and define

$$\Omega = \frac{2\mu}{\mu^2 - 1}, \quad \chi = \left(\frac{\mu^2+1}{\mu^2-1}\right) S,$$

which satisfy the first equation in (9.82). Using the definitions of R_1 and R_2 and (9.41), we can show that

$$\Omega = \frac{\alpha\beta(E - 3S^2 - C^2) - C\xi\eta}{2S^2(D\Lambda + 2\alpha\beta)}, \tag{9.84}$$

$$\chi = \frac{\Lambda^2 S^2 + D^2 - 4\alpha\beta C}{2(D\Lambda + 2\alpha\beta)}. \tag{9.85}$$

If we solve (9.84) for E we obtain (9.83). The equivalence of (9.81) and (9.85) can be established if we use (9.82) to substitute for S, C and D.

With a judicious choice of the rational quantities $\Theta, \Gamma, \Lambda, \Omega$ we can create solvable quintics with specific properties. For example, if we want $c = 0$, we take $\Theta = 0$ (see Example 9.3, in which $\Theta = 0, \Lambda = 0, \Gamma = 2, \Omega = -1$); for $d = 0$, take $\Gamma + \Lambda = 0$ (see Example 9.1, in which $\Theta = 1, \Lambda = -4, \Gamma = 4, \Omega = -1$). Also, Example 9.2 is $\Theta = 2, \Lambda = 2, \Gamma = 6, \Omega = 0$. To obtain a solvable Bring-Jerrard quintic (for which $c = d = 0$ and then $S \neq 0$ is guaranteed) we take $\Theta = 0$ and $\Gamma = -\Lambda$ so that

$$\chi = \Lambda^2/4, \quad S^2 = \frac{\chi^2}{w^2}, \quad E = S^2(3 + 4\Omega), \quad f = \Lambda(E - 25S^2)$$

and we recover the equations of Sect. 9.6 (see Examples 9.12 and 9.13).

In order to generate solvable quintics in which $S = 0$, $T \neq 0$, we scale χ, Θ, Γ as follows:

$$\chi = \chi_\Lambda/\Lambda, \quad \Theta = \Theta_\Lambda \Lambda, \quad \Gamma = \Gamma_\Lambda \Lambda$$

and then let $\Lambda \to \infty$ while ΛS remains finite. This leads to a choice of three rational numbers Θ_∞, Γ_∞, Ω and then

$$\chi_\infty = \frac{\Omega^2 - \Gamma_\infty^2 w^2}{4\Theta_\infty^3 w^2}, \quad C = \Theta_\infty \chi_\infty, \quad D = (\Gamma_\infty + 1)\chi_\infty,$$

$$E = C^2 - \frac{(T^2 - D^2)}{C}, \quad f = 2CD - \frac{D}{C^2}\left(T^2(1 - 2\Omega) - D^2\right),$$

where

$$T^2 = \chi_\infty^2/w^2.$$

(See Example 9.5, in which $\Theta_\infty = 1/2, \Gamma_\infty = 1, \Omega = -1$.) We cannot select $\Theta_\infty = 0$ or $\chi_\infty = 0$ and hence cannot generate a quintic for which $S = C = 0$ and $T \neq 0$ with these expressions. Since we cannot have $f = 0$, we cannot have $D = 0$ and hence cannot choose $\Gamma = -1$.

An alternative way of generating quintics for which $S = 0$ and $C \neq 0$, which does allow for the case $D = 0$, is as follows [27]. From (9.10), we have that when $S = 0$,

$$(C^2D + DE - Cf)^2 = T^2((5C^2 - E)^2 + 16CD^2),$$

where $T^2 = C^3 + D^2 - CE$. Assume $T \neq 0$ and let μ be such that

$$(5C^2 - E)^2 + 16CD^2 = 4\mu^2T^2.$$

Then we can show that

$$(5C^2 - E - 2\mu^2C)^2 = 4(\mu^2 - 4C)(D^2 + \mu^2C^2).$$

Now define λ via

$$\mu^2 - 4C = \lambda^2(\gamma^2 + \mu^2),$$

where we have introduced $\gamma = D/C$. We can think of μ, λ, γ as three arbitrary rational numbers and then C, D, E, f are in turn given by

$$\begin{aligned} 4C &= \mu^2 - \lambda^2(\gamma^2 + \mu^2), \quad D = \gamma C, \\ E &= 5C^2 - 2\mu^2C - 2C\lambda(\gamma^2 + \mu^2), \\ f &= \gamma(C^2 + E) - 2\mu(C^2 + C\gamma^2 - E). \end{aligned}$$

(See Example 9.6, in which $\mu = 1, \lambda = 3, \gamma = 0$.) The condition that $T \neq 0$ is $(\gamma^2 + \mu^2)(1 + \lambda)^2 \neq 0$.

We can generate a solvable quintic for which $S = T = 0$ by letting $\Omega \to \infty$ while Λ and ΩS remain finite. This results in

$$\begin{aligned} \chi &= \frac{\Lambda^2 - \Gamma^2}{4(\Theta + 1)\Theta^2}, \quad C = \Theta\chi, \quad D = (\Gamma + \Lambda)\chi, \\ E &= C^2 + \frac{D^2}{C}, \quad f = \frac{D^3}{C^2} + 2CD. \end{aligned} \tag{9.86}$$

(as in (9.65)). In this case we can dispense with χ altogether and just choose rational values of $C \neq 0$ and $D \neq 0$ and then calculate E and f from (9.86).

(See Example 9.8.) Note, however, that if we substitute for Θ and Γ in terms of C and D into the expression for χ, we find that

$$2\chi = \frac{4C^3 + D^2}{D\Lambda - 2C^2}$$

and hence we also cannot have $4C^3 + D^2 = 0$. If this condition is satisfied then we end up with a quintic with $\Delta = 0$; see (9.66).

To generate solvable quintics with $S^2 = C^2 \neq 0$ we can proceed more directly, choosing rational values for C, D and Λ. Taking $S = C$ first we note that in this case

$$a_2 = \xi = C\Lambda - D, \quad a_1 = -\eta = -C\Lambda - D.$$

From (9.40) we have

$$E = 4C^2 - u_1^3 u_2 - u_1 u_3^3 = 4C^2 + \eta \left(\frac{\xi}{-2C} + \frac{4C^2}{\xi} \right)$$

and from (9.28), (9.59) and (9.60),

$$f = -u_1^5 - u_2^5 - u_3^5 - 20C^2\Lambda = -\frac{\xi\eta^2}{4C^2} - \frac{2C\xi^2}{\eta} + \frac{16C^4\eta}{\xi^2} - 20C^2\Lambda.$$

(See Example 9.4.) We obtain exactly the same equations for E and f (in terms of C, D, Λ) if we assume $S = -C$. Given that we have chosen a specific sign for δ, which of $S = \pm C$ we get for a specific set C, D, Λ depends on the sign of C/ρ_S; see (9.58)).

9.8 Some Quintic Families

9.8.1 De Moivre

In the general theory, if we set $u_2 = u_3 = 0$ or $u_1 = u_4 = 0$, then we obtain a de Moivre quintic. De Moivre discovered this form while searching for equations that are satisfied by numbers of the form $\sqrt[n]{a + \sqrt{b}} + \sqrt[n]{a - \sqrt{b}}$; see [62].

We will assume that $u_2 = u_3 = 0$ with $C \neq 0$ (the case $u_1 = u_4 = 0$ is almost identical) and then, from (9.25)–(9.28),

$$2C = -u_1 u_4, \quad D = 0, \quad E = u_1^2 u_4^2, \quad f = -u_1^5 - u_4^5.$$

Since $E = 4C^2$, we have a quintic of the form

$$x^5 + 10Cx^3 + 20C^2x + f. \tag{9.87}$$

The discriminant of this is

$$\Delta = 5^5 \left(128C^5 + f^2\right)^2$$

and so, provided (9.87) is irreducible, the Galois group is F_{20}.

If a separable De Moivre quintic is reducible, then it can be shown that its Galois group is C_4. In turn, this means that if $\Delta \neq 0$, a reducible De Moivre quintic is the product of a linear and an irreducible quartic factor [8]. A family of such reducible De Moivre quintics is given by

$$x^5 + 5a^2x^3 + 5a^4x + 11a^5 = (x + a)(x^4 - ax^3 + 6a^2x^2 - 6a^3x + 11a^4).$$

The two quantities u_1^5, u_4^5 satisfy

$$u_1^5u_4^5 = -32C^5, \quad u_1^5 + u_4^5 = -f,$$

from which

$$u_1^5 = -\tfrac{1}{2}f + \sqrt{\tfrac{1}{4}f^2 + 32C^5}, \qquad u_4^5 = -\tfrac{1}{2}f - \sqrt{\tfrac{1}{4}f^2 + 32C^5},$$

with the fifth roots chosen so that $u_1u_4 = -2C$. If $(f^2/4) + 32C^5 > 0$, then u_1 and u_4 can both be taken as real. In this case x_1 is real, $x_2 = \overline{x_5}$, $x_3 = \overline{x_4}$, with a single real root. On the other hand, if $(f^2/4) + 32C^5 < 0$, then u_1 and u_4 are complex conjugates. In this case all the roots are real.

Example 9.14 Let $C = 1/2$, $f = 2$ so that

$$x^5 + 5x^3 + 5x + 2 = 0.$$

We find that

$$u_1 = \sqrt[5]{\sqrt{2} - 1}, \quad u_4 = -\sqrt[5]{\sqrt{2} + 1}$$

and the roots are therefore

$$x_{j+1} = \zeta^j \sqrt[5]{\sqrt{2} - 1} - \zeta^{4j} \sqrt[5]{\sqrt{2} + 1}, \quad j = 0, 1, 2, 3, 4.$$ ♦

As mentioned in Sect. 7.4, Galois proved that an irreducible quintic q (in fact a polynomial of any prime degree will do) is solvable by radicals if and only if all of the roots of q can be expressed as rational functions of any two of them. Explicit formulas to this effect for de Moivre quintics are given in [67].

9.8.2 Brioschi

Quintics of the form

$$x^5 - 10Zx^3 + 45Z^2x - Z^2 \tag{9.88}$$

are known as Brioschi quintics, named after the Italian mathematician Francesco Brioschi who, amongst many other achievements, founded the Polytechnic University of Milan in 1863. They form an important part of the general algorithms for solving quintic equations using elliptic functions and were studied extensively in the late nineteenth and early twentieth centuries. See Sects. 10.2 and 10.3.

Here we will assume that Z is rational and non-zero and determine under what conditions (9.88) is solvable in radicals. The method is analogous to that used in [64]. Note that a simple scaling will transform (9.88) into one whose constant term is any non-zero value we please. Thus if we write $Z = W^5/k^2$ and $x = \tilde{x}W^2/k$, (9.88) becomes

$$\frac{W^{10}}{k^5}\left(\tilde{x}^5 - 10W\tilde{x}^3 + 45W^2\tilde{x} - k\right).$$

The discriminant of (9.88) is

$$\Delta = 5^5 Z^8 (1728Z - 1)^2$$

and so any solvable Brioschi quintic has Galois group $G_q \cong F_{20}$. We construct the sextic resolvent (9.11) with root $y = \psi^2 = 500S^2$. Since Z is non-zero, we can replace y by $20wZ^2$ and remove the factor of $2^{12}5^6Z^{10}$ to yield a quadratic in Z:

$$\begin{aligned}(w^2 - 64w + 1600)(w^2 - 28w + 16)^2 Z^2 \\ - 2(5w^3 - 300w^2 + 4272w + 3200)Z - w + 25.\end{aligned} \tag{9.89}$$

There are no rational values of w which make the coefficient of Z^2 vanish and so a necessary and sufficient condition for there to be any rational solution pairs (w, Z) is that the discriminant of this quadratic in Z should be a square in $\mathbb{Q}$. That discriminant can be shown to be

$$4w(w - 20)^6.$$

We cannot have $w = 20$ since solving (9.89) shows that this corresponds to $Z = 1/1728$ and hence $\Delta = 0$. Thus we require that $w = r^2$, where $r \in \mathbb{Q}$. Then we can solve (9.89) and obtain the two roots

$$Z = \frac{5 \pm r}{(r^2 \mp 12r + 40)(r^2 \pm 6r + 4)^2}.$$

If we take the upper signs and construct the sextic resolvents for S and T we find that

$$S = -\frac{r(r+5)}{5P^2R}, \quad T = -\frac{(r+5)^2}{5P^3R},$$

where

$$P = r^2 + 6r + 4, \quad R = r^2 - 12r + 40 = (r-6)^2 + 4 > 0.$$

It is possible then to go through the solution process for general r and we find that

$$\begin{aligned} u_{1,4}^5 &= \frac{(r+5)^2\left(\pm\sqrt{5}r(2r^2 - 20r + 25) + 125(\pm 2\sqrt{5} - \sqrt{R})\right)}{250P^5R^{5/2}}, \\ u_{2,3}^5 &= \frac{(r+5)^4(25\sqrt{R} \pm (70 - 11r)\sqrt{5})}{250P^5R^{5/2}}. \end{aligned} \tag{9.90}$$

After the determination of u_1, we can then calculate u_2, u_3, u_4 from (9.39):

$$\begin{aligned} u_2 &= \frac{P\sqrt{R}}{2(r-5)^2}(5\sqrt{R} + (10 - r)\sqrt{5})u_1^2, \\ u_3 &= \frac{5P^2R}{2(r-5)^3(r+5)}(r(r + \sqrt{5R} - 10) + 50)u_1^3, \\ u_4 &= \frac{-P^3R^{3/2}}{8(r-5)^4(r+5)^2}(\sqrt{5}(3r - 10) + 5\sqrt{R})(\sqrt{5}(r-10) - 5\sqrt{R})^2u_1^4. \end{aligned} \tag{9.91}$$

Example 9.15 With $r = 0$ we have $Z = 1/128$ and hence the quintic

$$x^5 - \tfrac{5}{64}x^3 + \tfrac{45}{16{,}384}x - \tfrac{1}{16{,}384}.$$

The general formulas above show that

$$\begin{aligned} u_1^5 &= \tfrac{\sqrt{2}-2}{2^{18}}, \quad u_2 = 8(2 + \sqrt{2})u_1^2, \\ u_3 &= -2^7u_1^3, \quad u_4 = -2^{10}(2 + \sqrt{2})u_1^4. \end{aligned}$$

(continued)

Example 9.15 (continued)

If we write $x = \tilde{x}/8$, then the quintic becomes a multiple $(1/2^{15})$ of

$$\tilde{x}^5 - 5\tilde{x}^3 + \tfrac{45}{4}\tilde{x} - 2, \tag{9.92}$$

exhibiting the transformation described at the beginning of this section, with $W = 1/2$ and $k = 2$. The roots of (9.92) are obtained via $S = 0$, $T = -1$ and

$$u_{1,4}^5 = \tfrac{1}{8}(-2 \pm \sqrt{2}), \quad u_{2,3}^5 = \tfrac{1}{8}(10 \pm 7\sqrt{2}).$$ ♦

Example 9.16 Another interesting case is $r = 5$ ($Z = 2/59^2$) for which (9.91) is invalid. Then (9.90) shows that the reason for the problem is that $u_1 = 0$ and we can compute directly

$$u_2^5 = \frac{64}{59^5}, \quad u_3^5 = \frac{16}{59^5}, \quad u_4^5 = -\frac{4}{59^5}.$$

If we write $x = \tilde{x}/59$, then we are led to seek the roots of the quintic

$$\tilde{x}^5 - 20\tilde{x}^3 + 180\tilde{x} - 236.$$

The real root is given by

$$2u - u^2 + u^4, \quad u = \sqrt[5]{2}.$$ ♦

9.8.3 Quintics Derived from Elliptic Curves

Quintics of the form

$$y^5 + (a-3)y^4 + (b-a+3)y^3 + (a^2-a-1-2b)y^2 + by + a,$$

where $a, b \in \mathbb{Q}$, are known as Kondo-Brumer quintics and are solvable. They can be derived using properties of elliptic curves [43] and the Galois group of any such quintic is known to be isomorphic to D_{10} [39, Thm. 6], [43, Thm. 2.1]. To put this in depressed form we write $y = x - (a-3)/5$ and obtain

$$x^5 + \tfrac{1}{5}(7a - 2a^2 + 5b - 3)x^3 + \tfrac{1}{25}(4a^3 + 4a^2 - 7a - 5b - 15ab + 2)x^2$$
$$- \tfrac{1}{125}(3a-4)(a^3 + 11a^2 - 43a - 10b - 5ab + 3)x$$

$$+\tfrac{1}{3125}(4a^5+90a^4-815a^3+2020a^2+1670a$$
$$+300b+200ab-25a^2b-25a^3b-72).$$

The discriminant of both of these quintics is $\Delta = a^2h^2$, where

$$h=-a^2b^2-24a^3b+34a^2b+4a^5-4a^4-40a^3$$
$$+91a^2+30ab^2+14ab-4a+4b^3-b^2$$

and so $\delta \in \mathbb{Q}$. We will take $\delta = ah$, which is not necessarily positive. If we construct the sextic resolvents for ρ_S and ρ_T, we find that they always have rational solutions, which lead to

$$10\sqrt{5}S = 3a-b+1, \quad 50\sqrt{5}T = 8a^2-33a-7b-ab+2.$$

We can then show that

$$R_1^2 = \frac{5+\sqrt{5}}{6250}h, \quad R_2^2 = \frac{5-\sqrt{5}}{6250}h, \quad R_1R_2 = -\frac{h}{625\sqrt{5}}.$$

Continuing with general a and b leads to expressions that are exceedingly complicated and so we content ourselves with a specific example. The case $a=-2$ and $b=5$ is:

Example 9.17

$$y^5-5y^4+10y^3-5y^2+5y-2,$$
$$x^5+5x^2+10x+4,$$

with $y=x+1$. Applying the test (9.5) shows that there is one real root. We have $\delta = 1750$, $G_q \cong D_{10}$ and

$$S=\tfrac{1}{5}\sqrt{5}, \quad T=-\tfrac{3}{10}\sqrt{5},$$
$$R_1=\tfrac{1}{10}\sqrt{14(5-\sqrt{5})}, \quad R_2=\tfrac{1}{10}\sqrt{14(5+\sqrt{5})},$$
$$r_1=2, \quad r_2=-\tfrac{6}{5}\sqrt{5},$$
$$q_1=\tfrac{1}{5}\sqrt{70-\tfrac{154}{5}\sqrt{5}}, \quad q_2=-\tfrac{1}{5}\sqrt{70+\tfrac{154}{5}\sqrt{5}}.$$

Then

$$u_{1,4}^5=\tfrac{1}{10}\left(5-3\sqrt{5}\pm\sqrt{70-\tfrac{154}{5}\sqrt{5}}\right),$$
$$u_{2,3}^5=\tfrac{1}{50}\left(25+15\sqrt{5}\mp\sqrt{70(25+11\sqrt{5})}\right)$$

(continued)

Example 9.17 (continued)
and

$$u_2 = \tfrac{1}{4}\left(-5 + 3\sqrt{5} - \sqrt{70 - 14\sqrt{5}}\right) u_1^2,$$
$$u_3 = \tfrac{1}{4}\left(5 - 7\sqrt{5} - \sqrt{70(5 - \sqrt{5})}\right) u_1^3,$$
$$u_4 = \tfrac{1}{2}\left(25 - 15\sqrt{5} - \sqrt{70(25 - 11\sqrt{5})}\right) u_1^4.$$

♦

Two families of quintics that have Galois group C_5 are [43,44]

$$\begin{aligned} &y^5 + (m^5 - 3)y^4 \\ &\quad + (-m^9 - 2m^8 - 3m^7 - 5m^6 - 6m^5 - 2m^4 + m^3 - m^2 + 3)y^3 \\ &\quad + (m^{10} + 2m^9 + 4m^8 + 6m^7 + 10m^6 + 9m^5 + 4m^4 - 2m^3 + 2m^2 - 1)y^2 \\ &\quad - m^2(m^7 + 2m^6 + 3m^5 + 5m^4 + 5m^3 + 2m^2 - m + 1)y + m^5, \end{aligned} \tag{9.93}$$

with discriminant

$$\Delta = m^{18}(m^4 - 2m^3 + 4m^2 - 3m + 1)^4(m^4 + 3m^3 + 4m^2 + 2m + 1)^8$$

and

$$\begin{aligned} y^5 - n^2y^4 + 2(n^3 - 3n^2 + 5n - 5)y^3 - (n^4 - 5n^3 + 11n^2 - 15n + 5)y^2 \\ + (-n^3 + 4n^2 - 10n + 10)y - 1 \end{aligned} \tag{9.94}$$

with discriminant

$$\Delta = (n^3 - 5n^2 + 10n - 7)^2(n^4 - 5n^3 + 15n^2 - 25n + 25)^4.$$

For this second case we also have $S = 0$. Further examples of families of C_5 quintics associated with elliptic curves are given in [32].

As an example, if we take $m = -2$ in (9.93) we get

$$y^5 - 35y^4 + 215y^3 + 695y^2 + 180y - 32$$

and then the translation $y = x + 7$ leads to:

Example 9.18

$$x^5 - 275x^3 - 1650x^2 + 5500x + 41{,}800$$

Applying the test (9.5) shows that there are 5 real roots. We have $\delta = 968 \times 10^6$ and so the Galois group is either D_{10} or C_5. We can compute

$$\begin{gathered}
S = -\tfrac{11}{2}\sqrt{5}, \quad T = 55\sqrt{5}, \\
R_1 = 22\mathrm{i}\sqrt{50 + 10\sqrt{5}}, \quad R_2 = -22\mathrm{i}\sqrt{50 - 10\sqrt{5}}, \\
r_1 = -11{,}550, \quad r_2 = 4070\sqrt{5}, \\
q_1 = 110\mathrm{i}\sqrt{12{,}170 - 5378\sqrt{5}}, \quad q_2 = -110\mathrm{i}\sqrt{12{,}170 + 5378\sqrt{5}},
\end{gathered}$$

and the quartic resolvent is

$$\begin{aligned}
&\tfrac{1}{4}(2z^2 + (11,550 - 4070\sqrt{5})z + 100{,}656{,}875 - 44{,}289{,}025\sqrt{5}) \\
&\quad \times (2z^2 + (11,550 + 4070\sqrt{5})z + 100{,}656{,}875 + 44{,}289{,}025\sqrt{5}).
\end{aligned}$$

Then

$$\begin{aligned}
u_{1,4}^5 &= \tfrac{55}{2}\left(-105 + 37\sqrt{5} \pm 2\mathrm{i}\sqrt{12{,}170 - 5378\sqrt{5}}\right), \\
u_{2,3}^5 &= -\tfrac{55}{2}\left(105 + 37\sqrt{5} \pm 2\mathrm{i}\sqrt{12{,}170 + 5378\sqrt{5}}\right)
\end{aligned}$$

and

$$\begin{aligned}
u_2 &= \tfrac{1}{55}\left(5 - \mathrm{i}\sqrt{250 + 110\sqrt{5}}\right)u_1^2, \\
u_3 &= \tfrac{1}{242}\left(7 + \tfrac{17}{5}\sqrt{5} - 2\mathrm{i}\sqrt{58 + \tfrac{122}{5}\sqrt{5}}\right)u_1^3, \\
u_4 &= \tfrac{1}{6655}\left(-45 - 14\sqrt{5} - \mathrm{i}\sqrt{3650 + 1402\sqrt{5}}\right)u_1^4.
\end{aligned}$$

To determine the Galois group we need to apply one of the tests described in Sect. 9.5. This is an example where (9.72) does not immediately help since $W = 0$. However, if we solve the quadratic (9.70) we find that the solutions for Z^2 are 5500^2 and $176{,}000^2$ (and if we evaluate Z_1 from (9.71) we find

(continued)

Example 9.18 (continued)

that $Z_1 = 176{,}000$) and so $G_q \cong C_5$. This is confirmed by calculating ρ_1, ρ_2 and ρ_3 as defined in (9.68):

$$\rho_1 = 2^3 5^3 11^2 \times 1217, \quad \rho_2 = 2^3 5^2 11^2 \times 2689, \quad \rho_3 = 2^4 5^2 11^3 \times 19,$$

from which, see (9.69), $-5(\rho_1 + \rho_2 + 2\rho_3) = 2^{12} 5^4 11^2$, a square in $\mathbb{Q}$. ♦

If we take $n = 0$ in (9.94) we get:

Example 9.19

$$x^5 - 10x^3 - 5x^2 + 10x - 1$$

Applying the test (9.5) shows that there are 5 real roots. We have $\delta = 4375$ and so the Galois group is either D_{10} or C_5. We find that

$$S = 0, \quad T = -\tfrac{1}{2}\sqrt{5}, \quad r_1 = 1, \quad r_2 = \sqrt{5},$$
$$R_1 = \mathrm{i}\sqrt{\tfrac{1}{2}(5 - \sqrt{5})} = -q_1, \quad R_2 = -\mathrm{i}\sqrt{\tfrac{1}{2}(5 + \sqrt{5})} = -q_2$$

and the quartic resolvent is

$$\tfrac{1}{4}(2z^2 - (1 + \sqrt{5})z + 2)(2z^2 - (1 - \sqrt{5})z + 2).$$

This is an example where the solutions are most conveniently written in terms of complex exponentials noting, see Sect. 3.3, that $\mathrm{e}^{2\pi \mathrm{i}/n}$, where n is an integer, can always be written in terms of radicals. We find that

$$u_{1,4}^5 = \mathrm{e}^{\mp \mathrm{i}\pi/5} = \tfrac{1}{4}(1 + \sqrt{5}) \mp \tfrac{1}{4}\mathrm{i}\sqrt{10 - 2\sqrt{5}},$$
$$u_{2,3}^5 = \mathrm{e}^{\pm 3\mathrm{i}\pi/5} = \tfrac{1}{4}(1 - \sqrt{5}) \pm \tfrac{1}{4}\mathrm{i}\sqrt{10 + 2\sqrt{5}}$$

and appropriate fifth roots are

$$u_1 = \mathrm{e}^{-\pi \mathrm{i}/25}, \quad u_2 = \mathrm{e}^{-\pi \mathrm{i}/5} u_1^2 = \mathrm{e}^{-7\pi \mathrm{i}/25},$$
$$u_3 = \mathrm{e}^{2\pi \mathrm{i}/5} u_1^3 = \mathrm{e}^{7\pi \mathrm{i}/25}, \quad u_4 = \mathrm{e}^{\pi \mathrm{i}/5} u_1^4 = \mathrm{e}^{\pi \mathrm{i}/25}.$$

(continued)

Example 9.19 (continued)

With u_1 calculated as a fifth root from the formula above this gives all the roots in radicals using expressions already derived for $\exp(\pi\mathrm{i}/5)$ and $\zeta = \exp(2\pi\mathrm{i}/5)$. The roots of the quintic are

$$x_1 = 2\cos\tfrac{1}{25}\pi + 2\cos\tfrac{7}{25}\pi, \quad x_2 = 2\cos\tfrac{9}{25}\pi + 2\cos\tfrac{13}{25}\pi,$$

$$x_3 = 2\cos\tfrac{17}{25}\pi + 2\cos\tfrac{19}{25}\pi, \quad x_4 = 2\cos\tfrac{3}{25}\pi + 2\cos\tfrac{21}{25}\pi,$$

$$x_5 = 2\cos\tfrac{11}{25}\pi + 2\cos\tfrac{23}{25}\pi.$$

The Galois group is $G_q \cong C_5$ as can be seen from the fact that from (9.72) we find $W = 25$; the solutions for Z^2 in (9.70) are 25^2 and 175^2 and if we evaluate Z_1 from (9.71) we see that $Z_1 = 175$; ρ_1, ρ_2 and ρ_3 as defined in (9.68) are given by $-5/2$, $1/2$ and 1, respectively, and so $\rho_1 + \rho_2 + 2\rho_3 = 0$.

...

Another defining characteristic of quintics with Galois group C_5 is that all the roots can be expressed as rational functions of any given root [13, Thm. 13.2.6].

...Continuing Example 9.19. If one of the roots is x, then the others are (see [16])

$$\tfrac{1}{7}(-x^4 + 3x^3 + 8x^2 - 19x - 9), \quad \tfrac{1}{7}(2x^4 + x^3 - 23x^2 - 18x + 25),$$
$$\tfrac{1}{7}(3x^4 - 2x^3 - 24x^2 - 6x + 6), \quad \tfrac{1}{7}(-4x^4 - 2x^3 + 39x^2 + 36x - 22).$$

Some other examples of quintics with cyclic Galois group can be found in [16].

9.8.4 Quintics of the Form $x^5 + dx^2 + f$

We have seen that for Bring-Jerrard quintics there are an infinite number of essentially different solvable forms, corresponding to different values of Ω in Sect. 9.6. For trinomials of the form $x^5 + dx^2 + f$, the situation is fundamentally different [66]. If $q(x) = x^5 + dx^2 + f$ is irreducible and solvable, then there is a non-zero rational number s such that either

$$(d, f) = (5s^3, 3s^5),\ (5s^3, -15s^5) \text{ or } (25s^3, 300s^5), \qquad \text{with } G_q \cong D_{10}$$

or

$$(d, f) = (100s^3, 1000s^5) \text{ or } (250s^3, 625s^5), \qquad \text{with } G_q \cong F_{20}.$$

A proof of this statement is quite lengthy and so we will only give the briefest of outlines here. The roots of any of these quintics are simply s times the roots for the case $s = 1$ and so we only need consider this case. The starting point is the observation that if $C = E = 0$, the sextic (9.11) reduces to

$$(3125S^6 - 50DfS^2 - 16D^4)^2 = \frac{S^2\Delta}{3125}.$$

Then it can be shown that finding rational roots to this equation is equivalent to finding rational points on the elliptic curve

$$y^2 = x^3 + \tfrac{89}{100}x^2 + \tfrac{8}{25}x + \tfrac{1}{25}.$$

Below we give details of the five classes of solvable trinomials $x^5 + dx^2 + f$.

Example 9.20

$$x^5 + 5x^2 + 3,$$

$$G_q \cong D_{10}, \quad S = \tfrac{1}{5}\sqrt{5}, \quad T = -\tfrac{1}{10}\sqrt{5},$$

$$u_1^5 = \tfrac{1}{20}\left(-5 + \sqrt{5} + \sqrt{30 - \tfrac{66}{5}\sqrt{5}}\right),$$

$$u_2 = \tfrac{1}{4}\left(-5 + \sqrt{5} - \sqrt{6(5 + \sqrt{5})}\right) u_1^2,$$

$$u_3 = \tfrac{1}{4}\left(-5 + 3\sqrt{5} - \sqrt{30(5 - \sqrt{5})}\right) u_1^3,$$

$$u_4 = \tfrac{1}{4}\left(-25 + 5\sqrt{5} - \sqrt{150(5 - 2\sqrt{5})}\right) u_1^4. \qquad \blacklozenge$$

Example 9.21

$$x^5 + 5x^2 - 15,$$
$$G_q \cong D_{10}, \quad S = \tfrac{1}{5}\sqrt{5}, \quad T = \tfrac{1}{2}\sqrt{5},$$
$$u_1^5 = \tfrac{1}{100}\left(125 + 65\sqrt{5} - 7\sqrt{750 + 330\sqrt{5}}\right),$$
$$u_2 = \tfrac{1}{4}\left(-5 - \sqrt{5} - \sqrt{150 + 66\sqrt{5}}\right) u_1^2,$$
$$u_3 = \tfrac{1}{4}\left(25 + 3\sqrt{5} + 5\sqrt{6(5 + \sqrt{5})}\right) u_1^3,$$
$$u_4 = \tfrac{1}{4}\left(125 + 65\sqrt{5} + 7\sqrt{750 + 330\sqrt{5}}\right) u_1^4.$$
♦

Example 9.22

$$x^5 + 25x^2 + 300,$$
$$G_q \cong D_{10}, \quad S = \sqrt{5}, \quad T = -\tfrac{5}{2}\sqrt{5},$$
$$u_1^5 = -\tfrac{5}{2}\left(5 + \sqrt{5} + \sqrt{6(5 + \sqrt{5})}\right),$$
$$u_2 = \tfrac{1}{4}\left(-1 + \sqrt{5} - \tfrac{1}{5}\sqrt{5(30 + 6\sqrt{5})}\right) u_1^2,$$
$$u_3 = \tfrac{1}{20}\left(5 + 3\sqrt{5} - \sqrt{30(5 + \sqrt{5})}\right) u_1^3,$$
$$u_4 = \tfrac{1}{10}\left(-5 - \sqrt{5} + \tfrac{1}{5}\sqrt{5(30 + 6\sqrt{5})}\right) u_1^4.$$
♦

Example 9.23

$$x^5 + 100x^2 + 1000,$$
$$G_q \cong F_{20}, \quad S = 4, \quad T = -10,$$
$$u_1 = -2u, \quad u_2 = -2u^2, \quad u_3 = u^3, \quad u_4 = -u^4, \quad u = \sqrt[5]{2}.$$
♦

Example 9.24

$$x^5 + 250x^2 + 625,$$

$$G_q \cong F_{20}, \quad S = 5, \quad T = 0$$

$$u_1^5 = -25(-5 + 2\sqrt{5}),$$

$$u_2 = -\tfrac{1}{10}(5 + 3\sqrt{5})u_1^2, \quad u_3 = -\tfrac{1}{10}(1 + \sqrt{5})u_1^3,$$

$$u_4 = -\tfrac{1}{25}(5 + 2\sqrt{5})u_1^4.$$

♦

10 Epilogue

Our journey is over and we have reached our destination. Via a circuitous route, taking in complex numbers, fields, group theory and Galois theory, we have discovered that the generality and simplicity of the quadratic formula is the exception, not the rule. The formulas involved in solving cubic and quartic equations are hardly simple, but they are general. But that generality evaporates when we reach quintics, lost into the ether. Nevertheless, some quintic equations can be solved using only a finite number of the algebraic operations $+$, $-$, $\times$, $\div$ and $\sqrt{\ }$ and we have seen how to identify which quintics are solvable in this sense and shown how to solve them.

But .... (There is always a but!)

It was entirely natural to restrict the tools at our disposal to the basic arithmetic operations and the extraction of roots; after all this was sufficient for polynomial equations of degrees 2, 3 and 4. But can we enhance our toolbox with additional mathematical functions which then enable us to tame the quintic? The answer here is a resounding yes. In this chapter we will mention a few of these avenues, but be much less rigorous about including all the necessary background information. However, plenty of references are provided for readers who wish to follow up and find out more. Note that we are still interested in "formulas" for the roots, but loosening the restrictions on what can appear in those formulas. If one is solely interested in the numerical value of the roots of a polynomial, then there are a plethora of numerical techniques that can do the job efficiently.

10.1 Trigonometric Functions

Our first example goes back to the sixteenth century and was developed by Viète to avoid the need to navigate the troublesome waters posed by the newly fashioned complex numbers when solving cubic equations [38, Chapter 13]. Once again we

C. M. Linton, *Quintic Equations and How to Solve Them*,
https://doi.org/10.1007/978-3-032-01658-4_10

start from the depressed cubic $y^3 + py + q$, assuming that both p and q are real and that neither of them is zero, but now we make the substitution

$$y = \lambda \cos\theta$$

which yields

$$\lambda^3 \cos^3\theta + p\lambda\cos\theta + q.$$

The key observation is that we have the identity

$$\cos 3\theta = 4\cos^3\theta - 3\cos\theta,$$

valid for all θ, and so we can choose λ in order to make the combination of $\cos^3\theta$ and $\cos\theta$ in the cubic look like the right-hand side of this identity. The appropriate choice for λ is

$$\lambda = 2\sqrt{-p/3}$$

and then the cubic becomes

$$2\left(-\tfrac{1}{3}p\right)^{3/2}\left(4\cos^3\theta - 3\cos\theta\right) + q.$$

Thus, with the aid of the trigonometric identity, we see that the roots of the cubic correspond to values of θ for which

$$\cos 3\theta = \frac{3q}{2p}\sqrt{\frac{-3}{p}} = \sqrt{-\frac{q^2/4}{p^3/27}}. \tag{10.1}$$

We noted in Sect. 4.3 that the case of three real roots corresponds to a positive discriminant, i.e. $\Delta > 0$, where $\Delta = -4p^3 - 27q^2$. For this to be true, we must have $p < 0$ and $q^2/4 < -p^3/27$ so that the expression under the final square root symbol in (10.1) is a real number between 0 and 1 and we obtain a real value for 3θ in the interval $(0, \pi/2)$:

$$\theta = \tfrac{1}{3}\arccos\sqrt{-\frac{q^2/4}{p^3/27}}. \tag{10.2}$$

Since $\cos(x \pm 2\pi) = \cos x$, given a solution θ, then $\theta \pm 2\pi/3$ are also solutions and these three values for θ, when substituted into $y = 2\sqrt{-p/3}\cos\theta$ yield the three real roots of the cubic equation.

Example 10.1 Consider the equation

$$y^3 - 15y - 4 = 0,$$

which we looked at in Chap. 2, and for which $\Delta/108 = 121$. With the aid of Cardano's formula, we showed that the three solutions are 4, $-2 \pm \sqrt{3}$, but this required extracting the cube roots of the complex numbers $2 \pm 11\mathrm{i}$.

On the other hand, we have $\lambda = 2\sqrt{5}$ and, from (10.2), the three roots are given by

$$2\sqrt{5}\cos\theta, \quad 2\sqrt{5}\cos(\theta \pm 2\pi/3)$$

where

$$\theta = \tfrac{1}{3}\arccos\frac{2}{\sqrt{125}}.$$

The equivalence of the solutions is apparent from the fact that

$$2 \pm 11\mathrm{i} = \sqrt{125}\,\mathrm{e}^{\pm 3\mathrm{i}\theta}.$$ ♦

We can see if we can make the same idea work for quintics. The trigonometric identity which connects $\cos 5\theta$ with powers of $\cos\theta$ is

$$\cos 5\theta = 16\cos^5\theta - 20\cos^3\theta + 5\cos\theta \tag{10.3}$$

and so an immediate observation is that the same idea can only possibly work on quintics of the form

$$x^5 + 10Cx^3 + 5Ex + f.$$

With $x = \lambda\cos\theta$, this becomes

$$\frac{\lambda^5}{16}\left(16\cos^5\theta + \frac{160C}{\lambda^2}\cos^3\theta + \frac{80E}{\lambda^4}\cos\theta\right) + f$$

and so we will be able to use the trigonometric identity to good effect if both

$$\frac{160C}{\lambda^2} = -20 \quad \text{and} \quad \frac{80E}{\lambda^4} = 5$$

so that

$$\lambda^2 = -8C \quad \text{and} \quad \lambda^4 = 16E.$$

Immediately we see that this requires that $E = 4C^2$ and hence these are the De Moivre quintics

$$x^5 + 10Cx^3 + 20C^2x + f$$

considered in Sect. 9.8.1 and for which $\Delta = 5^5\left(128C^5 + f^2\right)^2$.

With $\lambda = 2\sqrt{-2C}$, we thus reduce the quintic to

$$8C^2\sqrt{-2C}\cos 5\theta + f$$

and the roots of the quintic can be found from solving

$$\cos 5\theta = -\frac{f}{8C^2\sqrt{-2C}} = -\sqrt{-\frac{f^2/4}{32C^5}}. \tag{10.4}$$

Looking back at Sect. 9.8.1, we see that the condition for five real roots for a De Moivre quintic is that $(f^2/4) + 32C^5 < 0$ and so in this case, where the algebraic solution requires taking fifth roots of complex quantities, the expression under the final square root symbol in (10.4) is again a real number between 0 and 1. However, $\cos 5\theta$ is negative and so there is a real value of 5θ in the interval $(\pi/2, \pi)$ which satisfies (10.4). The other four roots will be obtained by taking $\theta \pm 2\pi/5$ and $\theta \pm 4\pi/5$.

Example 10.2 Consider the equation

$$x^5 - 5x^3 + 5x + 2 = 0,$$

which corresponds to $C = -1/2$ and $f = 2$. This is a case for which $f^2/4 = -32C^5$ and so the discriminant is zero and there are multiple roots. We have $\lambda = 2$ and, from (10.4), the five roots are given by

$$2\cos\theta, \quad 2\cos(\theta \pm 2\pi/5), \quad 2\cos(\theta \pm 4\pi/5)$$

where $5\theta = \pi$. In other words, since $\cos(-\theta) = \cos\theta$, the roots are

$$-2, \quad 2\cos\tfrac{1}{5}\pi, \quad 2\cos\tfrac{1}{5}\pi, \quad 2\cos\tfrac{3}{5}\pi, \quad 2\cos\tfrac{3}{5}\pi.$$

(continued)

Example 10.2 (continued)

The quintic is reducible, one of the factors is $x + 2$ and since there are two double roots the accompanying quartic factor must be a perfect square. We find that

$$x^5 - 5x^3 + 5x + 2 = (x + 2)(x^2 - x - 1)^2.$$

If we solve the quintic using the method of Sect. 9.8.1 we can take $u_1 = u_4 = -1$ and $u_2 = u_3 = 0$, which yields the solutions $x_1 = -2$ and

$$x_2 = x_5 = -(\zeta + \zeta^4) = -2\cos\tfrac{2}{5}\pi = 2\cos\tfrac{3}{5}\pi,$$
$$x_3 = x_4 = -(\zeta^2 + \zeta^3) = -2\cos\tfrac{4}{5}\pi = 2\cos\tfrac{1}{5}\pi.$$ ♦

Example 10.3 Consider the equation

$$x^5 - 10x^3 + 20x + f = 0, \quad f^2 < 128,$$

which corresponds to $C = -1$ and, with the restriction on f, has five real roots. Note that changing f to $-f$ is equivalent to changing x to $-x$ and so the roots for negative f are just the negatives of the roots if f is replaced by $-f$. Hence we can assume that $f > 0$. We have $\lambda = 2\sqrt{2}$ and, from (10.4), the five roots are given by

$$2\sqrt{2}\cos(\pi - \tfrac{1}{5}\alpha), \quad 2\sqrt{2}\cos\tfrac{1}{5}(\pi \pm \alpha), \quad 2\sqrt{2}\cos\tfrac{1}{5}(3\pi \pm \alpha),$$

where

$$\alpha = \arccos(\tfrac{1}{16} f\sqrt{2}).$$

This achieves the goal of expressing the five real roots in terms of real quantities. With $f = 8$, we find that $\alpha = \pi/4$ and one of the roots is seen to be $x = 2$, leading to

$$x^5 - 10x^3 + 20x + 8 = (x - 2)(x^4 + 2x^3 - 6x^2 - 12x - 4).$$

If we solve the quintic using the method of Sect. 9.8.1 we have

$$u_1^5 = -\tfrac{1}{2}f + \sqrt{\tfrac{1}{4}f^2 - 32}, \quad u_4^5 = -\tfrac{1}{2}f - \sqrt{\tfrac{1}{4}f^2 - 32},$$

(continued)

Example 10.3 (continued)

with $u_1 u_4 = 2$ and $u_2 = u_3 = 0$. As expected, we have to take fifth roots of complex quantities in the case when $f^2 < 128$. When $f = 8$, we get

$$u_1 = \sqrt[5]{-4+4\mathrm{i}}, \quad u_4 = \sqrt[5]{-4-4\mathrm{i}}$$

and we can extract the roots by noting that $1 + \mathrm{i} = \sqrt{2}\,\mathrm{e}^{\mathrm{i}\pi/4}$. Thus, we find that

$$u_1 = -\sqrt{2}\,\mathrm{e}^{-\mathrm{i}\pi/20}, \quad u_4 = -\sqrt{2}\,\mathrm{e}^{\mathrm{i}\pi/20}$$

and then

$$\begin{aligned} x_1 &= -2\sqrt{2}\cos\tfrac{1}{20}\pi = 2\sqrt{2}\cos\tfrac{19}{20}\pi \\ x_2 &= \zeta u_1 + \zeta^4 u_4 \\ &= -\sqrt{2}(\mathrm{e}^{2\pi\mathrm{i}/5}\mathrm{e}^{-\mathrm{i}\pi/20} + \mathrm{e}^{8\pi\mathrm{i}/5}\mathrm{e}^{\mathrm{i}\pi/20}) = 2\sqrt{2}\cos\tfrac{13}{20}\pi \end{aligned}$$

and similarly for the other roots. In this way we obtain the same answers as previously. ♦

It is worth observing that trigonometric functions can be defined in terms of algebraic operations, its just that we need an infinite number of them. Thus, for example,

$$\cos x = 1 - \frac{x^2}{2!} + \frac{x^4}{4!} - \frac{x^6}{6!} + \ldots = \sum_{n=0}^{\infty}(-1)^n \frac{x^{2n}}{(2n)!}.$$

The infinite series on the right-hand side converges for all (real or complex) x, by virtue of the fact that the factorial function in the denominator grows more rapidly than any growth in the numerator. Once we allow infinite series into our set of potential solutions, a great many possibilities open up.

As one example, the equation

$$x^5 - x + f = 0$$

can be interpreted as setting $h(x) = f$, where $h(x) = x - x^5$. Lagrange's inversion theorem (see, for example, [12, §6.23]) then provides a mechanism for inverting the function h and deriving a series which expresses x as a function of f. It can be shown that

$$x = \sum_{n=0}^{\infty} \binom{5n}{n} \frac{f^{4n+1}}{4n+1},$$

in which $\binom{5n}{n}$ is a binomial coefficient. This series does not converge for all f, however. In fact it converges only for $|f| \leq 4/5^{5/4} \approx 0.535$ but for sufficiently small values of $|f|$ this series does give a root of the quintic, one that tends to zero as $f \to 0$. The failure of the series to converge when $|f|$ is greater than $4/5^{5/4}$ is directly related to the occurrence of a double root when $f^4 = 4^4/5^5$; see (9.73). The first person to explore this approach appears to have been Lambert in the 1750s and Eisenstein rediscovered it in the 1840s [50, Appendix], [53]. The method is extended in [28] to show that all the roots of $x^5 - x + f$ can be expressed in terms of hypergeometric functions and it is shown in [7] how to solve quintics with any of the forms $x^5 + ax^m + f = 0$, $m = 1, 2, 3$, in terms of hypergeometric functions.

10.2 Elliptic Functions

One of the spectacular successes of nineteenth-century mathematics was the development of a coherent theory of elliptic functions; these being functions of a complex variable which are doubly periodic. An elliptic function $f(z)$ thus satisfies

$$f(z + \omega_1) = f(z + \omega_2) = f(z)$$

for two different complex numbers ω_1 and ω_2, where it is assumed that ω_2 is not a real multiple of ω_1. The creation of this theory was largely achieved by Abel and Jacobi, though there were many other contributors, and the subject was developed extensively throughout the nineteenth century [57]. A thorough account of the history of elliptic and related functions throughout the nineteenth and early twentieth centuries can be found in [61]. This was a time when the implications of Galois' work were being gradually assimilated and it is unsurprising that connections between the theory of equations, the emerging theory of groups and the rich landscape of complex functions were a source of inspiration to many of the great mathematicians of the time.

One such was Charles Hermite, who had been interested in quintic equations from an early age. It is interesting to reflect that Hermite was in many ways the polar opposite of Galois: whereas Galois shunned detailed computations in favour of the creation of abstract structures, Hermite put calculations at the heart of his mathematics. Away from mathematics, Galois' revolutionary zeal contrasts vividly with Hermite's conservatism and support for the monarchy. In the 1850s, Hermite noticed that elliptic functions could be used to solve quintic equations of Bring-Jerrard type and thus, since the general quintic can be reduced to Bring-Jerrard form using algebraic operations, all quintics can be solved using such functions, together with root extractions. The general theory of elliptic functions and the transformation properties which are central to Hermite's discovery form an extensive body of work and we will not attempt to describe it here. Instead, we will sketch some ideas which hopefully will provide a glimpse into this world.

A modern treatment of this subject involves a menagerie of mathematical objects: elliptic integrals, elliptic functions, theta functions, modular functions, modular forms, all underpinned by a vast and powerful theory of functions of a complex variable. Here, we will use just one of these objects, elliptic integrals, to motivate the ideas that are involved. The **complete elliptic integral** of the first kind is defined via the integral

$$K(k) = \int_0^1 \frac{\mathrm{d}t}{\sqrt{(1-t^2)(1-k^2t^2)}}, \qquad 0 \le k < 1.$$

The number k is called the **modulus** of the elliptic integral and we have $K(0) = \pi/2$ and $K(k) \to \infty$ as $k \to 1$. This integral, and a number of other closely related ones, arises naturally in many applications; see, for example [41]. Perhaps the most common application is the determination of the period of oscillation of a simple pendulum. If we have a pendulum of length ℓ, then the often-quoted formula for its period is

$$T = 2\pi\sqrt{\ell/g}, \tag{10.5}$$

where g is the acceleration due to gravity. However, the derivation of this formula requires that we assume that the angle through which the pendulum swings is small. What if that is not the case? It turns out that if the maximum deviation of the pendulum from the vertical is through an angle α, then

$$T = 4\sqrt{\frac{\ell}{g}}\,K\left(\sin\tfrac{1}{2}\alpha\right),$$

which reduces to (10.5) if we set $\alpha = 0$ and is approximated well by (10.5) when α is small (the error is of order α^2).

A less common appearance of the function K, and one that is algebraic in nature, concerns the arithmetic-geometric mean (AGM) iteration that was introduced by Lagrange and worked on extensively by Gauss. The early history of the AGM iteration is detailed in [1] and [9] gives a very thorough analysis of the connections between the AGM iteration and elliptic functions. Suppose we start with two positive numbers a_0 and b_0 and then form a new pair by taking the arithmetic and geometric means of a_0 and b_0. In other words, let

$$a_1 = \tfrac{1}{2}(a_0 + b_0), \quad b_1 = \sqrt{a_0 b_0}.$$

We can then repeat this operation so that a_2 is the arithmetic mean of a_1 and b_1, whilst b_2 is their geometric mean, and so on. It is not difficult to show that $a_1 > a_2 > a_3 > \ldots$ and that $b_1 < b_2 < b_3 < \ldots$, with $a_n > b_n$. A little bit more effort reveals that in fact $a_n - b_n$ tends to zero as $n \to \infty$, so that the values of a_n decrease while those of b_n increase, to a common limit. This common limit is known as the arithmetic-geometric mean of a and b and denoted by $M(a, b)$. Gauss derived the remarkable result that

$$M(1+k, 1-k) = \frac{\pi}{2K(k)}. \tag{10.6}$$

It is easily seen that, for $\lambda > 0$,

$$M(\lambda a, \lambda b) = \lambda M(a, b)$$

and since $M(a_0, b_0) = M(a_1, b_1) = \ldots$, we also see that

$$M(a, b) = M\left(\tfrac{1}{2}(a+b), \sqrt{ab}\right). \tag{10.7}$$

If we combine these two results we can show that

$$M(1, b) = \tfrac{1}{2}(1+b)M\left(1, \frac{2\sqrt{b}}{1+b}\right). \tag{10.8}$$

Let us now explore the consequences of this result about arithmetic-geometric means for the elliptic integral K, given Gauss' result (10.6). It is customary when dealing with elliptic integrals to introduce the complementary modulus defined by

$$k' = \sqrt{1-k^2}$$

and to denote the value of K at k' by K', which in turn is thought of as a function of k. Thus $K'(k)$ is the same as $K(\sqrt{1-k^2})$. Since $k = \sqrt{1-k'^2}$, we can interchange k and k' in any formula. Combining (10.6) and (10.7) shows that

$$M(1, k') = \frac{\pi}{2K(k)}$$

and hence that

$$M\left(1, \frac{2\sqrt{b}}{1+b}\right) = \frac{\pi}{2K\left(\frac{1-b}{1+b}\right)}.$$

It follows from (10.8) with $b = k'$ that

$$K(k) = \frac{2}{1+k'} K\left(\frac{1-k'}{1+k'}\right) \tag{10.9}$$

and then if we replace k by $2\sqrt{k}/(1+k)$ (and therefore k' by $(1-k)/(1+k)$) in this formula, that

$$K\left(\frac{2\sqrt{k}}{1+k}\right) = (1+k)K(k).$$

This result is one of many forms of what is now known as Landen's transformation, John Landen being a somewhat obscure English mathematician who first derived the result in the 1770s. We also have

$$K'\left(\frac{2\sqrt{k}}{1+k}\right) = K\left(\frac{1-k}{1+k}\right) = \tfrac{1}{2}(1+k)K(k'),$$

the second equality coming from interchanging k and k' in (10.9).

If we now think of K'/K as a function of k in its own right, we see that we have demonstrated that

$$\frac{K'}{K}(k) = 2\frac{K'}{K}(l), \quad l = \frac{2\sqrt{k}}{1+k}. \tag{10.10}$$

To summarise; we have used the fact that the elliptic integral $K(k)$ can be expressed in terms of the arithmetic-geometric mean of $1+k$ and $1-k$, together with simple properties of the AGM, to show that the function K'/K evaluated at k is double that when evaluated at l, where k and l are related by an algebraic equation, namely

$$l^2(1+k)^2 - 4k = 0.$$

The AGM allows us to derive (10.10) relatively easily from (10.6), but it does not provide any clue to the fact that this is just an example of a more general phenomenon. It transpires that it is possible to find a relationship of the form

$$\frac{K'}{K}(k) = p\frac{K'}{K}(l), \tag{10.11}$$

for any prime p, with k and l related by an algebraic equation, known as a modular equation of degree p, and Jacobi derived these equations for $p = 3$ and $p = 5$. For $p = 3$, he showed that the modular equation is

$$\sqrt{kl} + \sqrt{k'l'} = 1$$

which is neatly expressed in terms of $u = k^{1/4}$ and $v = l^{1/4}$ as

$$u^4 - v^4 + 2uv(1 - u^2v^2) = 0. \tag{10.12}$$

For $p = 5$, with u and v as above, he found that

$$u^6 - v^6 + 5u^2v^2(u^2 - v^2) + 4uv(1 - u^4v^4) = 0. \tag{10.13}$$

It turns out that the Galois group of this equation (considered as an equation for v, given u) is none other than A_5. In general, when p is an odd prime, the modular equation is of degree $p + 1$ when expressed in terms of u and v and Galois established that, when $p = 5, 7$ or 11, it can be reduced to an equation of degree p, though he did not carry out the reduction. He stated (in the letter he sent to Chevalier the night before his duel in 1832; see [52, pp. 89–91]) that no such reduction was possible for higher values of p. The first proof of this appears to be that of Enrico Betti in 1853.

In the theory of elliptic functions, a quantity of fundamental importance is the **nome** $q(k)$ defined by

$$q = \mathrm{e}^{\pi i \tau}, \qquad \tau = \mathrm{i}\frac{K'}{K}$$

and we often work with q or τ as the independent variable, thus treating k as a function of q or τ rather than the other way around. Here we begin to see that in the general theory we are dealing with functions of a complex rather than a real variable. In fact, the theory of elliptic functions and a general theory for functions of a complex variable were developed in parallel.

The fundamental object of Hermite's approach is in fact the function $\phi(\tau)$, where $\phi^8 = k^2$, and it follows from the fact that the modular equation (10.13) ultimately stems from a relation of the form (10.11) with $p = 5$, that $u = \phi(\tau)$ and $v = \phi(5\tau)$ satisfy (10.13). There is a similarity here with the trigonometric identity (10.3) which expresses $\cos 5\theta$ in terms of $\cos\theta$. In that case the algebraic relationship between the quantities led to a de Moivre quintic, whereas in this case it can be shown that (10.13) can be reduced to a Bring-Jerrard quintic. The details are described in [9] as 'formidable'. Other sources are [49, §5.6], [2, §10.11], [29], [61, §5.1] and [31, §14.2].

The building blocks that form the basis of this reduction are the properties of the function $\phi(\tau)$ under the transformation

$$\tau \mapsto \frac{a + b\tau}{c + d\tau},$$

where a, b, c, d are integers and $ad - bc = 1$. There is an underlying group structure here and Galois had shown that, when $p = 5, 7$ or 11, that group structure enabled a reduction of the modular equation to one of lower degree. What Hermite set out to do, was to actually find an explicit form for that reduction in the case $p = 5$. He was able to show that, for a given $u = \phi(\tau)$, the solutions for v of the modular equation of degree 6 are

$$v_\infty = \phi(5\tau), \quad v_j = \phi((\tau + 16j)/5), \quad j = 0, 1, 2, 3, 4$$

and he then set

$$y = (v_\infty - v_0)(v_1 - v_4)(v_2 - v_3). \tag{10.14}$$

This then leads to a quintic equation for y which becomes

$$x^5 - x - A = 0, \quad A = \frac{2(1 + u^8)}{5^{5/4}u^2\sqrt{1 - u^8}}$$

if we write $y = 2 \cdot 5^{3/4}u\sqrt{1 - u^8}\, x$.

So in order to solve a general quintic we need to first reduce it to the form $x^5 - x - A$, which can be effected using radicals, and then find a value of τ for which

$$\frac{2(1 + u^8)}{5^{5/4}u^2\sqrt{1 - u^8}} = A.$$

If we square this equation we get a quartic in the variable $u^4 (= k(\tau))$, which again can be solved using radicals. Given $k(\tau)$, we can determine τ and then if we evaluate y at that value of τ using (10.14) we obtain a root of the quintic. The other roots follow from Galois' observations and are

$$\begin{aligned}
&(v_\infty - v_1)(v_2 - v_0)(v_3 - v_4),\\
&(v_\infty - v_2)(v_1 - v_3)(v_0 - v_4),\\
&(v_\infty - v_3)(v_2 - v_4)(v_1 - v_0),\\
&(v_\infty - v_4)(v_0 - v_3)(v_1 - v_2).
\end{aligned}$$

Alternative methods for solving quintic equations in terms of elliptic functions were developed around the same time by Leopold Kronecker and Francesco Brioschi. Kronecker's approach is summarised in [54] and a few details of Brioschi's contributions can be found in [30].

Hermite also showed how similar ideas could be used to solve the general quartic in terms of elliptic functions. This involves first transforming a given quartic into one of the form

$$x^4 - 6sx^2 - 8tx - 3s^2$$

and then associating the roots of this quartic with the solutions of the modular equation (10.12) for $p = 3$ via

$$u^8 = k^2 = 2s^{3/2}/(t + s^{3/2}).$$

As an algorithm for actually solving a quintic equation, Hermite's solution is totally impractical. This is primarily because the reduction of a given quintic to Bring-Jerrard form is usually exceedingly cumbersome. As a consequence, it was seen as desirable to devise solutions in terms of elliptic functions which start from a general principal quintic since the reduction to remove the x^4 and x^3 terms is considerably simpler than that required to obtain an equation with no x^2 terms either. This was achieved, independently, by Paul Gordan and Friedrich Kiepert in the 1870s. Both Kiepert's and Gordan's algorithms are described in [34]; see also [35, 36]. A rough outline of Kiepert's approach is as follows (Gordan's method is similar). First transform a given quintic into a principal quintic using a quadratic Tschirnhaus transformation, as described in Sect. 8.1. Next transform the principal quintic into a Brioschi quintic (see Sect. 9.8.2), again using a Tschirnhaus transformation that entails nothing worse than taking square roots. This transformation is described in [17, §120] and [24]. Equally, a Brioschi quintic can be transformed into a principal quintic [25] and this turns out to be somewhat simpler to use. Finally, Brioschi quintics possess a sextic resolvent which can be solved via elliptic functions, using ideas similar to those of Hermite.

10.3 The Icosahedron

In the latter half of the nineteenth century, attempts were made to link many branches of mathematics to geometry; there being a sense that in some way geometrical ideas are easier to visualise or are more intuitive, thus providing a better explanation of the reasons why such-and-such a result is true. The theory of the quintic equation is no exception, beginning with the ideas of Alfred Clebsch. Clebsch's work was subsequently eclipsed by Felix Klein, who related the solution of quintics to the properties of an icosahedron; see [42] and [31, Chap. 14]. Klein's *Vorlesungen über das Ikosaeder und die Auflösung der Gleichungen vom fünften Grade* (translated in [37]) has in many ways become the definitive work, providing a lens through which all previous contributions are seen. A modern approach to this subject can be found in [63].

An icosahedron is one of the five regular, or Platonic, solids. These are polyhedra which have the property that all the faces and all the vertices are identical, illustrated in Fig. 10.1, all of which were known to the ancient Greeks. Plato associated the icosahedron with water, the bulkiest of the four classical elements. A simple proof of the fact that there are only five such solids appears in Euclid's *Elements*. It relies on the fairly obvious observation that the sum of the interior angles of the faces which meet at the vertex of a regular polyhedron must be less than 360°. This condition is fulfilled only by three, four or five equilateral triangles (since the interior angles of

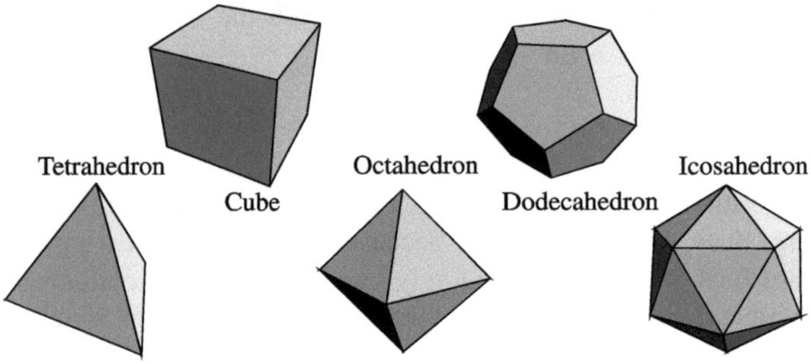

Fig. 10.1 The five Platonic solids

an equilateral triangle are 60° and $6 \times 60 = 360$), corresponding to a tetrahedron, octahedron and icosahedron, respectively, three squares (the interior angles of a square are 90° and $4 \times 90 = 360$), corresponding to a cube, and three pentagons (the interior angles of a pentagon are 108° and $4 \times 108 > 360$), corresponding to a dodecahedron. Three hexagons is already too many (the interior angles of a hexagon are 120° and $3 \times 120 = 360$).

The regular polyhedra can be characterised by the number of sides possessed by each face, p, and the number of faces meeting at each vertex, q, with $\{p, q\}$ known as the Schläfli symbol for the polyhedron. The icosahedron has 20 triangular faces, so $p = 3$, and five faces meet at each of its 12 vertices, so $q = 5$; the Schläfli symbol for the icosahedron is thus $\{3, 5\}$. The vertices are connected via 30 edges.

Let G be the rotational symmetry group of a regular polyhedron with Schläfli symbol $\{p, q\}$. If we consider a vertex A of the polyhedron (centred at O) then the symmetries which fix A are the rotations about OA through $2\pi n/q, n = 0, \ldots, q - 1$. These symmetries form a subgroup of G known as the stabilizer of A, or stab(A). We thus have $|\operatorname{stab}(A)| = q$. On the other hand, the orbit of A under the action of G is simply the set of vertices of the polyhedron. It therefore follows from a basic result in group theory known as the orbit-stabilizer theorem, that

$$|G| = qV,$$

where V is the number of vertices. A similar consideration of symmetries which fix the midpoint of an edge or the centroid of a face show that

$$|G| = 2E, \quad \text{and} \quad |G| = pF,$$

with E the number of edges and F the number of faces. This shows that the group of rotational symmetries of an icosahedron is of order 60 (we will refer to this group as I_{60}) and a little more work shows that this group is isomorphic to the alternating group, A_5. If we include reflections, we find that the full symmetry group for the icosahedron is isomorphic to S_5 and hence the group of symmetries of an icosahedron and the Galois group of a general quintic are essentially the same.

In order to convert geometrical properties of a regular polyhedron into algebraic statements so as to see how far we can push this correspondence, we need to introduce a few new concepts. The first is **stereographic projection** which allows us to associate complex numbers with points on a sphere. Consider a sphere of unit radius centred at the origin O of some rectangular coordinate system (X, Y, Z) and label the point on the sphere directly above O, $(0, 0, 1)$, as the north pole, N, and that directly below O, $(0, 0, -1)$, as the south pole, S. For any point P on the sphere, other than the north pole, we can draw a line from N through P and, if necessary, extend it so that it intersects the X, Y-plane at some point $(x, y, 0)$. We then associate the complex number $z = x + iy$ with the point P. Thus, for example, S corresponds to $z = 0$.

As P approaches the north pole, $|z|$ gets larger and larger and it is natural to associate a single point ∞ with N. The addition of the point ∞ to $\mathbb{C}$ produces the so-called **extended complex plane**:

$$\hat{\mathbb{C}} = \mathbb{C} \cup \{\infty\}.$$

With the addition of the correspondence between N and ∞ this provides a bijection between the unit sphere and $\hat{\mathbb{C}}$. The extended complex plane interpreted in this way is usually referred to as the **Riemann sphere**.

If we inscribe a regular polyhedron inside a unit sphere (so that each of the vertices lies on the sphere) then we can associate a complex number (or ∞) with each point P' on the polyhedron by first extending the line OP' so that it intersects the sphere at P (for a vertex, P' and P are the same) and then using stereographic projection to associate an element of the extended complex plane to our original point.

Instead of representing a complex number z in terms of two real numbers ($z = x + iy$) it can be convenient to represent it as the ratio of two complex numbers, known as **homogeneous coordinates**,

$$z = \frac{u}{v}, \qquad u, v \text{ not both zero.}$$

If $v = 0$ but $u \neq 0$, then z is taken as the point at infinity. In this way any pair of complex numbers, not both zero, generates a unique element of the extended complex plane $\hat{\mathbb{C}}$. On the other hand, for any non-zero complex number λ, the pairs (u, v) and $(\lambda u, \lambda v)$ give rise to the same z. The great advantage of this way of thinking is that, when considering the extended plane, ∞ no longer needs to be treated as a special case.

A homogeneous function of two complex variables u and v (of degree d) is a function F which satisfies

$$F(\lambda u, \lambda v) = \lambda^d F(u, v)$$

for all $\lambda \in \mathbb{C}$. If F is a homogeneous polynomial in u and v then it is often simply referred to as a form. A homogeneous polynomial of degree 1 is a linear form. A homogeneous polynomial of degree 2 is known as a quadratic form. Any polynomial equation of degree n in the variable x can be written in terms of a form by writing $x = u/v$ and then multiplying through by v^n. Hence $a_n x^n + a_{n-1}x^{n-1} + \cdots + a_1 x + a_0 = 0$ is equivalent to

$$a_n u^n + a_{n-1}u^{n-1}v + \cdots + a_1 u v^{n-1} + a_0 v^n = 0.$$

We now turn to the question of how geometrical rotations can be expressed in terms of maps on the extended complex plane. It can be shown that any rotation of the unit sphere is equivalent to a map on the Riemann sphere given by

$$z \mapsto \frac{az + b}{-\overline{b}z + \overline{a}}, \tag{10.15}$$

for some pair of complex numbers a and b that satisfy $|a|^2 + |b|^2 = 1$. See, for example, [17, §122], [51, p. 289], or [69, §18]. Note that if we replace a by $-a$ and b by $-b$ we get the same transformation. In particular, a rotation through χ around the polar axis ON is equivalent to $b = 0$ and $a = \pm e^{i\chi/2}$, or

$$z \mapsto e^{i\chi} z.$$

In terms of homogeneous coordinates, (10.15) can be expressed as the pair of transformations

$$u \mapsto au + bv, \qquad v \mapsto -\overline{b}u + \overline{a}v.$$

A regular icosahedron has 12 vertices and we can arrange the icosahedron so that one of the vertices is at the north pole of the unit sphere, as illustrated in Fig. 10.2. The vertices are N, S, A_j, B_j, $j = 1, 2, 3, 4, 5$ and we can choose the orientation of the (X, Y, Z) coordinate system so that A_3 and B_5 lie in the plane $Y = 0$. Rotations around ON through angles $2j\pi/5$, $j = 0, 1, 2, 3, 4$, are symmetries and we can represent these as rotations on the Riemann sphere via the transformations $z \mapsto \zeta^j z$, where $\zeta = e^{2\pi i/5}$ is a fifth root of unity. We will label the rotation $z \mapsto \zeta z$, a rotation through $2\pi/5$ around ON, by R.

Another symmetry, which we will denote by T, is a rotation through an angle π around OP, where P is the midpoint of A_3N (we could have chosen any of the points A_j; our choice of coordinates means that A_3 is convenient for calculations). It can be shown that suitable combinations of R and T carry the vertex at N to any of the 12 vertices of the icosahedron. The only symmetries which leave N fixed are the five distinct powers of R and hence R and T generate the $5 \times 12 = 60$ rotational symmetries of the icosahedron. Hence

$$I_{60} = \langle R, T \rangle, \quad R^5 = T^2 = id.$$

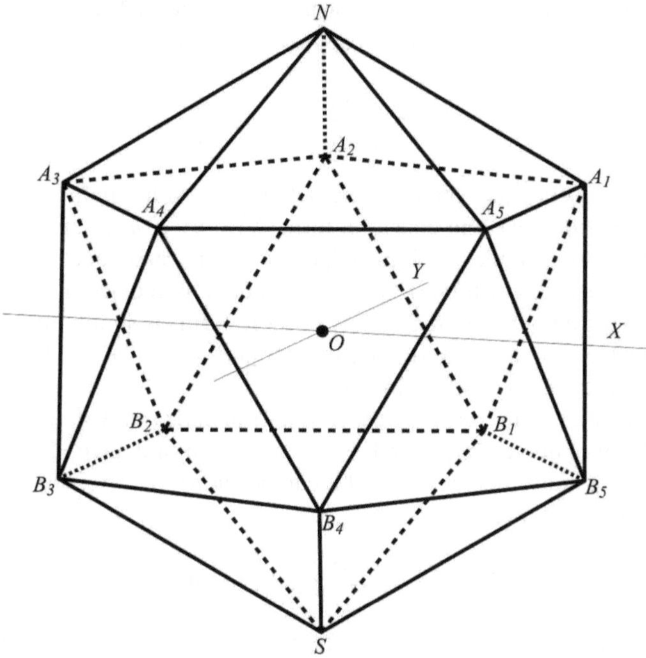

Fig. 10.2 An icosahedron with vertices on the unit circle. The (X, Y, Z) coordinates of N are $(0, 0, 1)$ while those of S are $(0, 0, -1)$. With $\theta = \arctan 2$ and $\phi = \pi/5$, the coordinates of the points A_n are $(\sin\theta\cos(2n-1)\phi, \sin\theta\sin(2n-1)\phi, \cos\theta)$ and those of B_n are $(\sin\theta\cos 2n\phi, \sin\theta\sin 2n\phi, -\cos\theta)$

It can be shown that T is equivalent to the map

$$z \mapsto \frac{pz+1}{z-p},$$

where because of our choice of coordinates, p is real and negative. In fact

$$p = \frac{-1}{\zeta+\zeta^4} = \zeta^2+\zeta^3 = -\cos\tfrac{1}{5}\pi = -\tfrac{1}{2}(1+\sqrt{5}).$$

Combining R and T then shows that the 12 vertices correspond to the points

$$0,\ \infty,\ \zeta^j(\zeta^2+\zeta^3),\ \zeta^j(\zeta+\zeta^4),\quad j = 0, 1, 2, 3, 4 \tag{10.16}$$

on the Riemann sphere.

For a given regular polyhedron with Schläfli symbol $\{p, q\}$ we can try and find a rational function $f : \hat{\mathbb{C}} \to \hat{\mathbb{C}}$ which is invariant under the action of G and we can seek to do this by finding a pair of homogeneous polynomials (of the same degree) which are themselves unaffected by an application of any element of G:

$$f(z) = \frac{P(u, v)}{Q(u, v)}, \quad z = \frac{u}{v}.$$

To do this we associate any point on the polyhedron with a point on the Riemann sphere as described above and construct forms $\mathcal{V}(u, v)$, $\mathcal{E}(u, v)$ and $\mathcal{F}(u, v)$ which vanish, respectively, at the vertices, edge midpoints, and face centroids of the polyhedron. The zeroes of these forms are unaltered by the action of G and so the effect of applying an element of G to any of these forms is simply to multiply them by a constant factor α and it turns out that we must have $|\alpha| = 1$. Any form for which $\alpha = 1$ is called an absolute invariant of G.

Since $\mathcal{V}$ vanishes at the vertices, it will be a form of degree $V = |G|/q$. Similarly $\mathcal{E}$ will be a form of degree $|G|/2$ and $\mathcal{F}$ will be a form of degree $|G|/p$. The forms $\mathcal{V}^q$, $\mathcal{E}^2$ and $\mathcal{F}^p$ all have degree $|G|$ and it can be shown [69, p. 336] that they must be linearly dependent, i.e.

$$c_1\mathcal{V}^q + c_1\mathcal{E}^2 + c_3\mathcal{F}^p = 0,$$

for some complex numbers c_1, c_2, c_3, not all zero. It can also be shown that any form which is invariant under the action of G is a polynomial in the forms $\mathcal{V}$, $\mathcal{E}$ and $\mathcal{F}$.

A form $\mathcal{V}(u, v)$ that vanishes at the vertices of the icosahedron listed in (10.16) is the product

$$\mathcal{V}(u, v) = uv \prod_{j=0}^{4} \left(u - \zeta^j pv\right) \prod_{k=0}^{4} \left(u + \zeta^k v/p\right)$$

which can be shown to reduce to

$$\mathcal{V} = uv\left(u^{10} + 11u^5v^5 - v^{10}\right).$$

It is also possible to construct forms $\mathcal{F}(u, v)$ and $\mathcal{E}(u, v)$ which vanish at, respectively, the face centroids and edge midpoints of the icosahedron, though we will omit any of the details. We obtain

$$\mathcal{F} = -(u^{20} + v^{20}) + 228(u^{15}v^5 - u^5v^{15}) - 494u^{10}v^{10}$$

and

$$\mathcal{E} = u^{30} + v^{30} + 522(u^{25}v^5 - u^5v^{25}) - 10{,}005(u^{20}v^{10} + u^{10}v^{20}).$$

The linear dependence of $\mathcal{V}$, $\mathcal{F}$ and $\mathcal{E}$, which are all absolute invariants of I_{60}, is given by

$$1728\mathcal{V}^5 - \mathcal{E}^2 - \mathcal{F}^3 = 0. \tag{10.17}$$

The **icosahedral equation** is what we get when we try and solve for z one of $\mathcal{V}^5 = w\mathcal{E}^2$, $\mathcal{V}^5 = w\mathcal{F}^3$ or $\mathcal{E}^2 = w\mathcal{F}^3$, with w given, which are all essentially the same as a consequence of (10.17). Taking the first of the three options results in

$$z^5\left(z^{10} + 11z^5 - 1\right)^5 = w\left(z^{30} + 1 + 522(z^{25} - z^5) - 10{,}005(z^{20} + z^{10})\right)^2,$$

which is an equation of degree 12 in z^5. This equation cannot be solved in radicals but it can be solved in terms of elliptic functions, or hypergeometric functions [37, Part I, Chap. V, §7], [69, §25], and [50].

To complete the story we now need to relate the icosahedral equation to a quintic. We have seen that the rotational symmetries of the icosahedron form a group I_{60} that is isomorphic to A_5, the group of even permutations of five letters. Can we identify five geometrical objects that play the role of these letters? One way to do this is to note that an icosahedron has 30 edges and an octahedron has 6 vertices. It is thus possible to inscribe an octahedron inside an icosahedron, with the vertices of the octahedron placed at the midpoints of the edges of the icosahedron, in five different ways.

In exactly the same way that we constructed the form $\mathcal{V}$ which vanishes at the vertices of an icosahedron, we can construct a form $\mathcal{O}_j$, $j = 1, 2, 3, 4, 5$, for each of the inscribed octahedra, which vanishes at its vertices. Details can be found in [17, §131] or [63, Chap. 4.8]. For a given z on the Riemann sphere, it can be shown that the five numbers $x_j = \mathcal{O}_j(z)$ are the roots of the Brioschi quintic (see Sect. 9.8.2)

$$x^5 - 10wx^3 + 45w^2x - w^2$$

in which

$$w = \mathcal{V}^5/\mathcal{E}^2,$$

which is just the icosahedral equation. We noted in Sect. 10.2 that it is possible to transform between a principal and a Brioschi quintic using fairly simple transformations that require nothing more than square roots. Thus a general quintic can be transformed via a principal quintic into a Brioschi quintic with parameter w. If we can then solve the icosahedral equation $\mathcal{V}^5 = w\mathcal{E}^2$ for z, we can compute the roots of the Brioschi quintic $x_j = \mathcal{O}_j(z)$ and then invert the transformations to recover the roots of our original quintic. The connections between quintic equations and the symmetries of an icosahedron are used in [18] and [14] as the basis for iterative solutions of the quintic.

It was Galois who first understood that the key to unlocking the secrets of polynomial equations was inextricably linked to the concept of symmetry. In algebraic terms, and specifically with reference to the roots of a polynomial, a symmetry can be thought of as some property of those roots that is invariant when the order of the roots is changed. Symmetries such as these can often be described geometrically and the deep connections discovered by Klein connecting the roots of a quintic equation with the symmetries of an icosahedron are perhaps the ultimate manifestation of this.

References

1. Almkvist, G., Berndt, B.: Gauss, Landen, Ramanujan, the arithmetic-geometric mean, ellipses, π, and the Ladies Diary. Am. Math. Month. **95**(7), 585–608 (1988)
2. Armitage, J.V., Eberlein, W.F.: Elliptic Functions. London Mathematical Society Student Texts, vol. 67 (Cambridge University Press, Cambridge, 2006)
3. Ayoub, R.G.: Paolo Ruffini's contributions to the quintic. Arch. History Exact Sci. **23**, 253–277 (1980)
4. Berndt, B.C., Spearman, B.K., Williams, K.S.: Commentary on an unpublished lecture by G. N. Watson on solving the quintic. Math. Intell. **24**(4), 15–33 (2002). https://doi.org/10.1007/BF03025320
5. Berwick, W.E.H.: The condition that a quintic equation should be soluble by radicals. Proc. Lond. Math. Soc. **2**(1), 301–307 (1915)
6. Bewersdorff, J.: Galois Theory for Beginners. A Historial Perspective. American Mathematical Society, Providence (2006). Translated by David Kramer
7. Birkeland, R.: On the solution of quintic equations. In: Proceedings of the ICM, Toronto 1924. University of Toronto Press (1928)
8. Borger, R.L.: On DeMoivre's quintic. Am. Math. Month. **15**(10), 171–174 (1908)
9. Borwein, J.M., Borwein, P.B.: Pi and the AGM. A Study in Analytic Number Theory and Computational Complexity. Wiley, Hoboken (1987)
10. Cohen, H.: A Course in Computational Algebraic Number Theory. Springer, Berlin (1993)
11. Cooke, R.: Classical Algebra. Its Nature, Origins and Uses. Wiley, Hoboken (2008)
12. Copson, E.T.: An Introduction to the Theory of Functions of a Complex Variable. Clarendon Press, Oxford (1935)
13. Cox, D.A.: Galois Theory, 2nd edn. Wiley, Hoboken (2012)
14. Crass, S.: Solving the quintic by iteration in three dimensions. Exp. Math. **10**(1), 1–24 (2001). https://doi.org/10.1080/10586458.2001.10504425
15. Crilly, T., Weintraub, S.H., Wolfson, P.R.: Arthur Cayley, Robert Harley and the quintic equation: newly discovered letters 1859—1863. Historia Math. **44**, 150–169 (2017). https://doi.org/10.1016/j.hm.2016.09.003
16. Diaz y Diaz, F.: A table of totally real quintic number fields. Math. Comput. **56**(194), 801–808 (1991)
17. Dickson, L.E.: Modern Algebraic Theories. B. H. Sanborn, Chicago (1926)
18. Doyle, P., McMullen, C.: Solving the quintic by iteration. Acta Math. **163**, 151–180 (1989)
19. Dummit, D.S.: Solving solvable quintics. Math. Comput. **57**(195), 387–401 (1991)
20. Fauvel, J., Gray, J. (eds.): The History of Mathematics: A Reader. Macmillan Press, New York (1987)
21. Fine, B., Rosenberger, G.: The Fundamental Theorem of Algebra. Springer, Berlin (1997)
22. Gårding, L.: Mathematics and Mathematicians. Mathematics in Sweden Before 1950. History of Mathematics, vol. 13. American Mathematical Society, Rhode Island (1998)
23. Garver, R.: The Tschirnhaus transformation. Ann. Math. **29**(1–4), 319–333 (1927)

C. M. Linton, *Quintic Equations and How to Solve Them*,
https://doi.org/10.1007/978-3-032-01658-4

24. Garver, R.: On the Brioschi normal quintic. Ann. Math. **30**(1–4), 607–612 (1928)
25. Garver, R.: On the transformation which leads from the Brioschi quintic to a general principal quintic. Bullet. Am. Math. Soc. **36**(2), 115–119 (1930)
26. Glashan, J.C.: Notes on the quintic. Am. J. Math. **7**(2), 178–179 (1885)
27. Glashan, J.C.: The isodyadic quintic. Am. J. Math. **45**(4), 251–254 (1923)
28. Glasser, M.L.: Hypergeometric functions and the trinomial equation. J. Comput. Appl. Math. **118**, 169–173 (2000). https://doi.org/10.1016/S0377-0427(00)00287-9
29. Goldstein, C.: Charles Hermite's stroll through the Galois fields. Rev. d'histoire Math. **17**, 211–270 (2011)
30. Gray, J.: Linear differential equations and group theory from Riemann to Poincaré. Springer, New York (1986)
31. Gray, J.: A History of Abstract Algebra. From Algebraic Equations to Modern Algebra. Springer, Berlin (2018)
32. Hashimoto, K., Tsunogai, H.: Generic polynomials over Q with two parameters for the transitive groups of degree five. Proc. Japan Acad. Ser. A **79**, 142–145 (2003). https://doi.org/10.3792/pjaa.79.142
33. Kiernan, B.M.: The development of Galois theory from Lagrange to Artin. Arch. History Exact Sci. **8**(1–2), 40–154 (1971)
34. King, R.B.: Beyond the Quartic Equation. Birkhäuser, Basel (1996)
35. King, R.B., Canfield, E.R.: An algorithm for calculating the roots of a general quintic equation from its coefficients. J. Math. Phys. **32**(4), 823–825 (1991). https://doi.org/10.1063/1.529338
36. King, R.B., Canfield, E.R.: Icosahedral symmetry and the quintic equation. Comput. Math. Appl. **24**(3), 13–28 (1992). https://doi.org/10.1016/0898-1221(92)90210-9
37. Klein, F.: Lectures on the Icosahedron and the Solution of the Fifth Degree. Cosimo, New York (2007). Originally published in 1884. Translated by G. G. Morrice
38. Kline, M.: Mathematical Thought from Ancient to Modern Times. Oxford University Press, Oxford (1972)
39. Kondo, T.: Some examples of unramified extensions over quadratic fields. Sci. Rep. Tokyo Women's Christian Univ. **120–121**, 1399–1410 (1997)
40. Lavallee, M.J., Spearman, B.K., Williams, K.S.: Watson's method of solving a quintic equation. JP J. Algebra Number Theory Appl. **5**(1), 49–73 (2005)
41. Lawden, D.F.: Elliptic Functions and Applications. Springer, Berlin (1989)
42. Lê, F.: Alfred Clebsch's "Geometrical Clothing" of the theory of the quintic equation. Arch. Hist. Exact Sci. **71**, 39–70 (2017). https://doi.org/10.1007/s00407-016-0180-5
43. Lecacheux, O.: Constructions de polynômes génériques à groupe de Galois résoluble. Acta Arithmet. **86**(3), 207–216 (1998)
44. Lehmer, E.: Connection between Gaussian periods and cyclic units. Math. Comput. **50**(182), 535–541 (1988)
45. McClenon, R.B.: A contribution of Leibniz to the history of complex numbers. Am. Math. Month. **30**(7), 369–374 (1923)
46. McClintock, E.: On the resolution of equations of the fifth degree. Am. J. Math. **6**(1), 301–315 (1883–1884)
47. McClintock, E.: Analysis of quintic equations. Am. J. Math. **8**(1), 45–84 (1885)
48. McClintock, E.: Further researches in the theory of quintic equations. Am. J. Math. **20**(2), 157–192 (1898)
49. McKean, H., Moll, V.: Elliptic Curves. Function Theory, Geometry, Arithmetic. Cambridge University Press, Cambridge (1999)
50. Nash, O.: On Klein's icosahedral solution of the quintic. Expo. Math. **32**, 99–120 (2014). https://doi.org/10.1016/j.exmath.2013.09.003
51. Needham, T.: Visual Complex Analysis. Clarendon Press, Oxford (1997)
52. Neumann, P.M.: The mathematical writings of Évariste Galois. European Mathematical Society, Helsinki (2011)
53. Patterson, S.J.: Eisenstein and the quintic equation. Historia Math. **17**, 132–140 (1990)

54. Petri, B., Schappacher, N.: From Abel to Kronecker. Episodes from 19th century algebra. In: Laudal, O.A., Piene, R. (eds.) The Legacy of Niels Henrick Abel, pp. 227–266. Springer, Berlin (2004)
55. Prasolov, V.V.: Polynomials. Springer, Berlin (2004)
56. Rabinowitz, S.: The factorization of $x^5 \pm x + n$. Math. Mag. **61**(3), 191–193 (1988)
57. Rice, A.: In search of the "birthday" of elliptic functions. Math. Intell. **30**(2), 48–56 (2008)
58. Roland, G., Yui, N., Zagier, D.: A parametric family of quintic polynomials with Galois group D_5. J. Number Theory **15**(1), 137–142 (1982). https://doi.org/10.1016/0022-314X(82)90087-7
59. Rosen, M.I.: Neils Hendrik Abel and equations of the fifth degree. Am. Math. Month. **102**(6), 495–505 (1995)
60. Rothman, T.: Genius and biographers: the fictionalization of Evariste Galois. Am. Math. Month. **89**(2), 84–106 (1982)
61. Roy, R.: Elliptic and Modular Functions. From Gauss to Dedekind to Hecke. Cambridge University Press, Cambridge (2017)
62. Schneider, I.: Der Mathematiker Abraham de Moivre (1667–1754). Arch. History Exact Sci. **5**, 177–317 (1968)
63. Shurman, J.M.: Geometry of the Quintic. Wiley, Hoboken (1997)
64. Spearman, B.K.: Solvable Brioschi resolvents. Appl. Algebra Eng. Commun. Comput. **13**, 447–452 (2003). https://doi.org/10.1007/s00200-002-0108-y
65. Spearman, B.K., Williams, K.S.: Characterization of solvable quintics $x^5 + ax + b$. Am. Math. Month. **101**(10), 986–992 (1994). https://doi.org/10.1080/00029890.1994.12004578
66. Spearman, B.K., Williams, K.S.: On solvable quintics $x^5 + ax + b$ and $x^5 + ax^2 + b$. Rocky Mount. J. Math. **26**(2), 753–772 (1996)
67. Spearman, B.K., Williams, K.S.: DeMoivre's quintic and a theorem of Galois. Far Eastern J. Math. Sci. **1**(1), 137–143 (1999)
68. Toscano, F.: The Secret Formula. How a Mathematical Duel Inflamed Renaissance Italy and Uncovered the Cubic Equation. Princeton University Press, Princeton (2020). Translated by A. Sangalli
69. Toth, G.: Glimpses of Algebra and Geometry, 2nd edn. Springer, New York (2002)
70. Yang, L.: Recent advances on determining the number of real roots of parametric polynomials. J. Symbol. Comput. **28**, 225–242 (1999). https://doi.org/10.1006/jsco.1998.0274
71. Yang, L., Hou, X.R., Zheng, Z.B.: A complete discrimination system for polynomials. Sci. China Ser. E Technol. Sci. **39**(6), 628–646 (1996)
72. Young, G.P.: Resolution of solvable equations of the fifth degree. Amer. J. Math. **6**(1), 103–114 (1883–1884)
73. Young, G.P.: Solution of solvable irreducible quintic equations, without the aid of a resolvent sextic. Am. J. Math. **7**(2), 170–177 (1885)
74. Young, G.P.: Solvable quintic equations with commensurable coefficients. Am. J. Math. **10**(2), 99–130 (1888)

Index

C. M. Linton, *Quintic Equations and How to Solve Them*,
https://doi.org/10.1007/978-3-032-01658-4

MIX
Papier aus verantwortungsvollen Quellen
Paper from responsible sources
FSC® C105338

If you have any concerns about our products,
you can contact us on
ProductSafety@springernature.com

In case Publisher is established outside the EU,
the EU authorized representative is:
Springer Nature Customer Service Center GmbH
Europaplatz 3, 69115 Heidelberg, Germany

Printed by Libri Plureos GmbH
in Hamburg, Germany